RISK MANAGEMENT AND SYSTEM SAFETY

by
LEONAM DOS SANTOS GUIMARÃES

Frontier India

An imprint of
Frontier India Technology
No 22, 4th Floor, MK Joshi Building, Devi Chowk, Shastri Nagar,
Dombivli West, Maharashtra, India. 421202
http://frontierindia.org
https://www.facebook.com/frontierindiapublishing
The views expressed in this book are those of the author and not of the publisher. The publisher is not responsible for the views of the author and authenticity of the data, in any way whatsoever. Cataloging / listing of this book for resale purpose can be done, only by authorized companies. Cataloging / listing or sale by unauthorized distributors, bookshops, booksellers etc., is strictly prohibited and will be legally prosecuted. All disputes are subject to Thane, Maharashtra jurisdiction only.

Contents

Index of images, charts and tables

... Ed ptarid... N ...

1. Definitions

1.1 Systems General Theory

1.1.1 Characteristics of a System

A system is a certain set of discrete elements (or components) interconnected or in interaction [1]. The word "certain" means that the considered system can be identified, which is a prerequisite for it to be analyzed. It is also necessary to notice that the definition indicates that the system is made of component parts in interaction: thus it is assumed that the system is not simply the sum of its parts. This implies that if the nature of a component part changes as a result of a certain event, the entire system is changed.

In Literature, it is possible to find different words to designate the hierarchy among the component parts of a system. With the purpose of standardization and clarity, in this work we will use the notions of item, component, subsystem, elementary system and system, with the following inclusive relations.

$$(\text{ITEM}) \subset (\text{COMPONENT}) \subset (\text{SUBSYSTEM}) \subset$$
$$(\text{ELEMENTARY SYSTEM}) \subset (\text{SYSTEM})$$

This way, a system will be represented by Image 1.1.1-A. It is characterized by:

a) the choice of a resolution limit defining the considered components, which will define the level of detail of the analysis to be performed on the system; the boundaries among various systems will be called "interface";

b) the choice of the outer limits and the system environment; the outer limits envelope the set of interacting systems with the studied system and are essential to the description of the system itself and its functions.

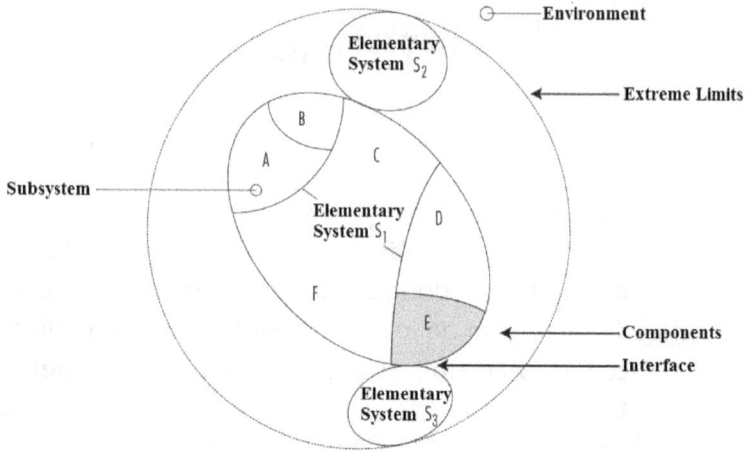

Image 1.1.1 - A) Representation of a System

In this work, the term "entity" will be used always when there is no need to make reference to the internal structure of the system (therefore to its component parts). Otherwise, the term "system" will be used.

Every system is generally defined by one or more functions (or tasks) that must be met under conditions and within a pre-established environment (required performance), from certain component parts.

The most important characteristics of a system that must be specified before analysis are:

a) system functions:
- main functions or tasks;
- secondary functions or tasks; and
- hierarchy (order of importance) of the functions.

b) system structure:
- component parts, with its individual functions, its functioning and performance characteristics;
- interrelations of component parts; and
- localization of component parts.

c) functioning conditions of the system:
- functioning states
- functioning conditions of the components and the system; and
- possibilities of configuration changes (or reconfigurations).

d) operating conditions (or utilization, in a wider sense)
- system monitoring (or surveillance, in a wider sense) conditions (alarms, inspections, verifications, periodic tests);
- intervention conditions over the system (maintenance, repair); and
- technical specifications of operation, in other words, the procedures, conditions, restrictions that must be respected during the operation of the system.

e) system environment:
- other elementary systems of the facility or the global process within which the analyzed system is located;
- set of human operators that intervene over the system; and
- environment per say, that can be manifested by unfavorable ambient conditions (temperature, pressure, humidity, vibration, levels of electromagnetic and ionizing radiation), particularly severe weather phenomena (floods, winds, sea states), or external aggressions originated by nature (earthquakes, tsunamis, volcanism) or by industry (aircraft crashes, explosions nearby).

During the different phases of the system design, not all of these characteristics are known. This makes approximations and hypotheses necessary. As the system design moves forward and this different information can be specified, the operational safety analysis should be corrected, modified, improved, in other words, should be periodically reviewed.

1.1.2 System Analysis

A system analysis is a process guided towards acquisition, investigation and ordered treatment of specific information to the system and relevant to a decision or a goal. This process implies the development of a system model.

According to this definition, the system analysis main function is acquiring information. This process must be done according to rules or methods, otherwise the corresponding model can prove to be little useful or unsuited to the objectives.

The first question to be set is: what information should be obtained? The answer is so evident as it may look at first sight.

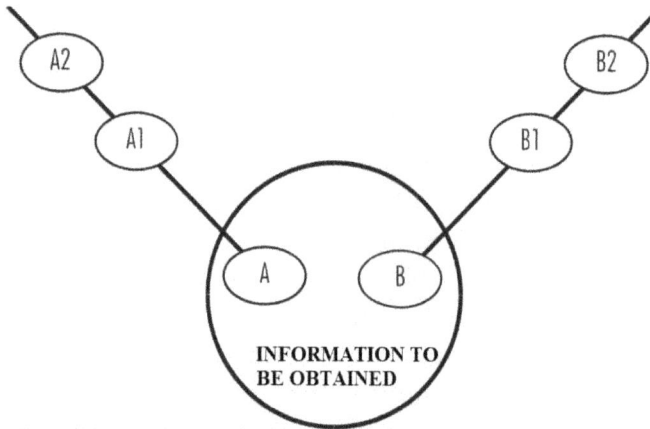

Image 1.1.2 - A) Required Information to the System Analysis

Consider Image 1.1.2-A. A circle represents the information that is essential to be obtained for the system analysis objective. An expert analyst in type A problems begins his study in this area and is led to some interesting questions that lead to the A1 area. The search for the A1 area leads him to A2, and so on. Another expert analyst in type B problems follows the same procedure that takes to B1, then B2, and so on.

To picture this problem, consider a safety system of a large industrial facility: the analyst starts with the failure causes of electrical nature of the system drives, goes on examining the system internal electrical sources, then the external power supplies. When the period of work is over, especially the time to present the analysis or to make decisions, the relevant information is not available, despite the great efforts made.

In order to avoid this "drift" in the analysis, it is very important to establish, already from the beginning, the system main characteristics to be considered, such as:

a) inner boundaries analysis: where will be notably set the physical, geographical and functional system boundaries, as well as interfaces to other systems and the environment; and

b) resolution boundaries analysis: where will be notably set the analysis depth levels, in other words, if it should fall to the level of item, or be confined to the level of component or subsystem or elementary system.

Clearly, these limits can be reviewed throughout the study, but this review should be done with full knowledge of all its implications, such as workload, time analysis, weight distribution efforts by the most important areas.

The process of analysis show lead to a first system model. Additional studies lead to new information that are aggregated to the model, generating reviews that must converge for the final model of the system. The analysis conclusion, as well as the decisions to be made will so be based on this final model. The Image 1.1.2-B shows this concept.

REAL SYSTEM

- **acquisition**
- investigation
- information
  processing

1st MODEL

- **acquisition**
- investigation
- supplementary
  information
  processing

FINAL MODEL

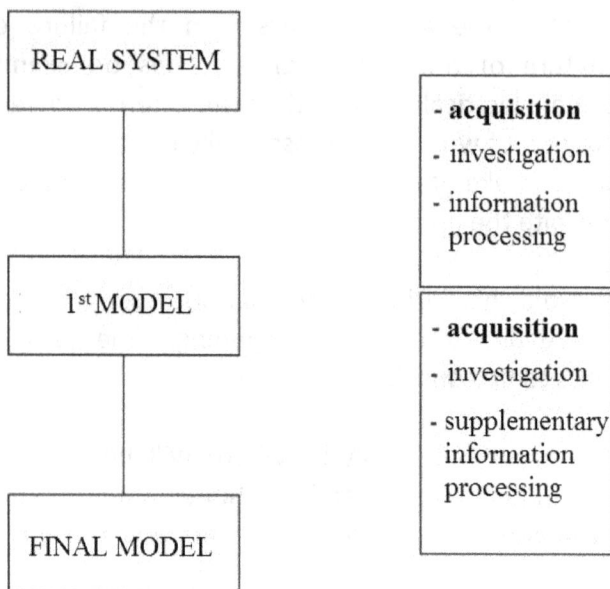

Image 1.1.2-B) Stages of the Process of Analysis

1.1.3 Failure

Failure is understood as an event that interrupts the ability of an entity to meet a required function [2].

One can say that an entity has failed when it ceases to be able to fulfill its duties. By extension, it is possible to consider that there is a failure when there is a change in the ability of an entity to fulfill a required function: the associated tolerances, in this case, must be set.

With the goal to better set this notion of failure, several classifications can be established [2]:

a) by speed of occurrence:

- progressive failure: failure due to an evolution in time of the characteristics of an entity; generally it can be anticipated by surveillance (monitoring, inspection, tests, rehearsal); and
- abrupt failure: failure that does not happen by a progressive loss of performance and that cannot be anticipated by surveillance.

b) by amplitude:
- partial failure: it results from diversion of characteristics beyond specified boundaries, but does not imply such a way that interrupts completely the required function;
- complete failure: it results from diversion of characteristics beyond specified boundaries, implying a complete interruption of the required function.

c) by conjugate speed x amplitude:
- cataleptic failure: complete and abrupt failure at the same time
- degradation failure: progressive and partial failure at the same time, that in a long-term can become complete.

d) by birth time within the system life: consider a set of supposed identical entities and in operation since the initial time t = 0. Define the failure rate as the proportion λ, normalized per unit of time, of entities that, having survived at any instant t, suffer a complete failure until the time t + Δ. Passing to the limit, we obtain the rate of instantan failure, which is a function of t. Note that the entities often have a failure rate versus time as a said curve "bathtub" (Image 1.1.3-A).

Three distinct periods are deduced from this curve, associated to three different types of failure rate, on the useful life of the entities that have this characteristic:

- early failure (or juvenile), which occurs at the beginning of the entity life, with a decreasing rate of occurrence over time, the beginning of life being counted from a specific time t = 0.

- constant failure rate, which occurs in most part of the entity life in a substantially constant way; and
- wear failure, which appears at the end of the entity life, with an increasing rate of occurrence over time, usually due to aging processes related to the entity, such as deterioration due to fatigue and corrosion.

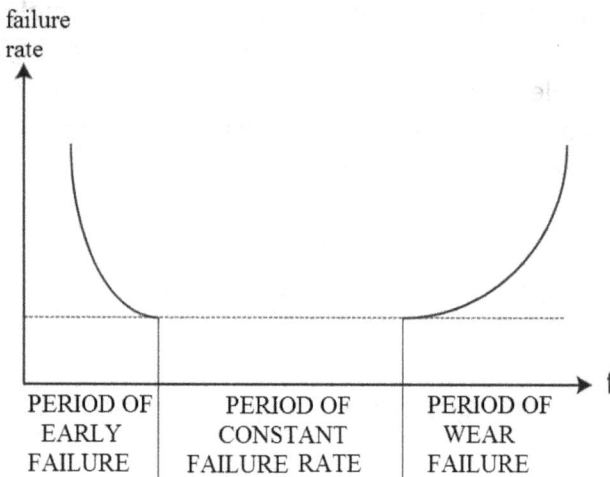

Image 1.1.3-A) "Bathtub" Curve

e) by effects:

Failures that occur in a system are likely to cause very different effects. Certain failures do not directly affect the system functions and require only a simple corrective action (non-urgent repair, for instance). Other failures affect the system availability or its safety. The failure effects are then assessed, using a gravity scale through which the system operating degradation levels are distributed in categories or classes. Usually we consider the four categories in Table 1.1.3-B.

MINOR FAILURE	Failure that disrupts the proper operation of a system, causing negligible damage to the said system or its environment without presenting risks to human life
SIGNIFICANT FAILURE	Failure that disrupts the proper operation of a system, without causing significant damage or presenting significant risks to human life
CRITICAL FAILURE	Failure that implies the interruption of one or more essential functions of a system and causes serious damage to the said system or its environment, presenting, however, small risks of death or serious injuries
CATASTROPHIC FAILURE	Failure that causes the loss of one or more essential functions of a system, causing serious damage to said system or its environment and / or implies death or serious injury

Table 1.1.3-B) Classification of Failures by Effects

This classification can equally qualify the consequences of an event. However, there is the following difference between the two concepts.

- effects are all manifestations resulting from the occurrence of an event, which is assumed to be produced in isolation;
- consequences are all logical sequences associated with incidence of an event; they thus comprehend the effects of the event when it occurs, for example, at the same time of another event; to some extent, one can speak of minor, significant, critical, catastrophic consequences.

f) by causes:

The entity failure results from the cause of failures, which are defined as "circumstances related to the design, manufacture or use and that imply failure." Therefore causes have failure as consequence by a failure mode, defined as any "physical, chemical process, or other that implies a failure".

Failures can then be classified into three categories according to their causes:

- primary failure: failure of an entity whose direct or indirect cause is not associated with another entity; usually it is necessary to repair the entity to put it back into operation; if we considered a component, the cause will be internal, that is, an item; it may be due to wear problems, design or manufacturing defects, operating technical specification defects; for example, a pipe that after pressurization on a lower pressure than which it was sized for, breaks;
- secondary failure: failure of an entity whose direct or indirect cause is the failure of another entity and for which that entity was not qualified and sized; usually it is necessary to repair the entity to put it back into operation; other entities failures, particular environmental conditions, human error can lead to secondary failure; for example, a pipe breaks after being pressurized to a higher pressure than which it was sized for, as a result from the failure of another component;
- command failure: failure of an entity whose direct or indirect cause is the failure of another entity for which this entity was qualified and sized; usually is not necessary to repair the entity to put it back into operation; a failure such as this occurs when an entity changes state as an effect of a unintended control signal (inadvertent closing of a valve, for example).

Table 1.1.3-C. represents these tree failure categories. A classification such as this should be considered as a guide to the analyst. It will be seen that it will be very useful to the cause tree method application.

"Aging"	Respect to Dimensioning	Primary Failure
Design	Human	
Manufacturing	Error	
Installation		
Human Error of Operation	Excessive Conditions and Loading	Secondary Failure
Environment		
Other Components		
Other Components	Incorrect command and control signals	Command Failure
Environment		
Human Error of Operation		

Table 1.1.3-C) Classification of Failures by Causes

1.1.4 Breakdown

A very close concept to failure is the concept of breakdown: "failure of an entity to fulfill a required function". After an entity fails, it is said that it had a breakdown, in other words, a breakdown always results from a failure. Clearly, breakdowns can be classified similarly to failures. However, there is a special breakdown classification related to the possibilities of its confirmation [2]:

a. intermittent breakdown: an entity breakdown that occurs during a limited period of time, after which the entity returns to fulfill its required functions, without being submitted to a corrective maintenance operation;

b. elusive breakdown: an entity breakdown that is intermittent and hardly observed;

c. permanent breakdown: an entity breakdown that remains until a corrective maintenance operation is done;

    d. latent breakdown: existent breakdown that has not yet been observed; it is also named "hidden breakdown";

1.1.5 Relations among Defect, Failure and Breakdown

Defect is every deviation between a characteristic of an entity and a required characteristic, such deviation exceeds limits of acceptability, within given conditions [2]. This way, a defect is defined as a non-compliance to the objectives or specifications.

However, not all defects lead to a failure: an observed defect in the system or on a system component may well not affect the ability of the system to perform a required function.

In contrast, all failures lead to a breakdown state, which characterizes a defect, because the interruption of the entity ability to fulfill its function is indisputably linked to a deviation. A failure must then be characterized by its immediate effects and the breakdown by the entity state in the longer term.

Image 1.2.1-B shows the relation among defect, failure and breakdown.

1.2 Failure Modes

1.2.1 Conception

Having set the concepts of failure and failure cause, it becomes important to set the concept of failure mode: "a failure mode is the effect in which a failure is observed" [2].

This way, at every failure of an entity, modes and causes are associated: modes are generated by causes, a mode representing the effect or effects for which manifests the cause. Image 1. 2. 1-B illustrates the close links and interrelations that exist between these concepts.

This image also illustrates the difficulty for sometimes there is to differentiate the modes and causes of failure. To help in this distinction, it can be said that the component causes of failures are usually component items failures. The failure modes are, in turn, the translation of the effects of these failures on the component functions.

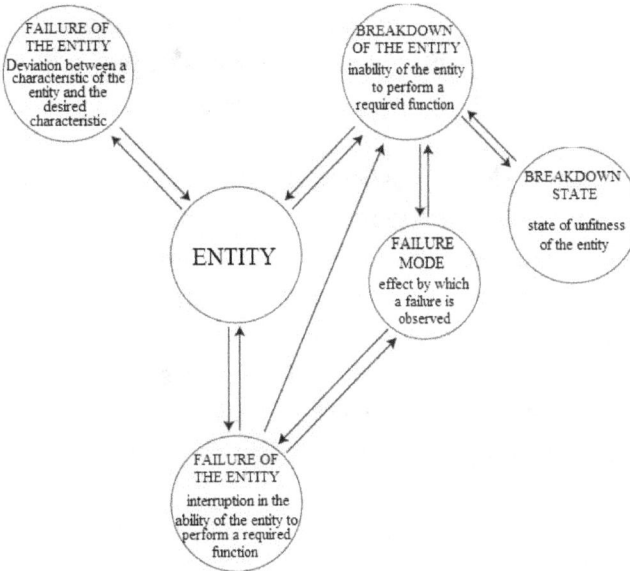

Image 1.2.1-A) Defect, Failure and Breakdown of a System

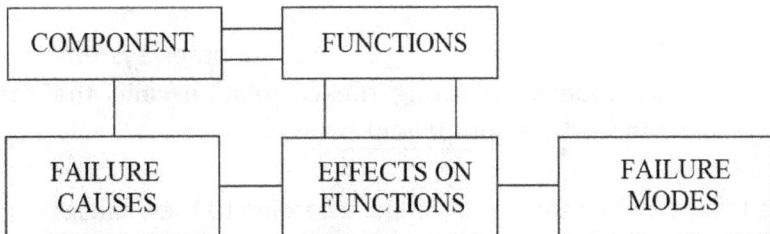

Image 1.2.1-B) Causes, Effects and Modes of Failure

So, all failure causes must be associated with failure modes: if a failure cause cannot be linked to failure modes, for instance, because its effects on the functions are different from those previously considered, it is necessary to draw up a new failure mode.

The concepts of causes and modes failures are deeply connected to the system decay levels: so, for example, failure modes of a system component can be the failure causes of this system, or the causes of a failure mode defined at the system level.

The failure mode is closely associated with the model adopted for the failure effects. It also depends on the component functions within the system. The failure modes may differ, for the same type of component, according to their function within different systems. So, for a pump, a lower flow to nominal will be a failure mode of this pump within a particular system, but in case of another system it may be necessary to consider several lower flow possibilities to nominal with various failure modes associated.

Failure modes can be classified according to their spread:

a) Independent Failure Mode: one of strictly random nature, that individually occurs to an entity, and is therefore not the result of the spread of failure modes of other entities of the same system; and

b) Dependent Failure Mode: one whose occurrence is conditional upon the occurrence of other system entities failure modes, featuring the correlations and the spread among multiple component failures.

The failure modes can be classified according to their causes:

a) Simple Cause Failure Mode: characterized as the effect of an internal or external initiator event occurrence to the considered entity, that only affects the entity;

b) Common Cause Failure Mode: characterized as the effect of an internal or external initiator event occurrence to the considered entity, which simultaneously affects other systems components entities.

1.2.2 Dependency among failures

Dependent failures are those which occur simultaneously or concomitantly over multiple entities and that hold among them dependent relations. This type of failure arises with the increasing complexity of industrial systems and ever greater reliability and safety requirements, leading to component redundancy.

Redundancy is the multiplication of the necessary components to the system so that it fulfills its mission, thus overcoming one or more failures. However, failures which have the same origin (design errors, for example) can simultaneously affect redundant components, therefore compromising the effectiveness of redundancy.

The treatment of dependent or common cause failure modes leads to particularly difficult problem of analysis. Its approach is however essential, since most recorded accidents, especially for complex systems, in the practical reality is almost always processed under these modes.

Two failures E_1 and E_2 are dependent if and only if [3]:
$$P[E_1.E_2]= P[E_1]*P[E_2|E_1] \neq P[E_1]*P[E_2]$$

In other words, the probability of the event $E_1.E_2$ cannot be expressed only as the product of the probabilities of events E_1 and E_2. In general, given a set of failures E_1, E_2,..., E_n, if these failures are dependent [3]:
$$P[E_1.E_2. \ldots .E_i. \ldots .E_n]= P[E_1]*P[E_2|E_1]* \ldots P[E_i|E_1.E_2. \ldots E_{i-1}]* \ldots *P[E_n|E_1.E_2. \ldots . E_{n-1}]$$

So, the probability of each successive failure depends on failures that precede the sequence. It is shown that the probability of dependent failures is greater than the product of the probabilities of these supposed independent failures [3].

The definition of dependent failure demands some commentaries:

a) Occurrence level of failures: they affect multiple entities (items, components, subsystems, elementary systems); the components can belong to the same elementary system or to different elementary systems; dependent failures of the "human operator" system are also considered;

b) Simultaneity or concomitance: the failures appear simultaneously, in other words, at the same time, or in a concomitant way, that is, in a sequence that takes place within a certain period of time; if this period of time is too large, they can be confused with multiple independent failures;

c) Dependency relations: they translate the physical, functional, human or other connections that establish a determinist cause-effect relationship among failures.

Dependent failures can be classified into three categories [4]:

a) initializing events that cause common cause failures: comprehend internal and external events to the system that could potentially cause an accident due to the induction of multiple failures of elementary systems; these events often have serious consequences on the facility, its components and its structures; to name as examples fires, floods, earthquakes, aircraft falls, for external events, and loss of electrical power or of an important elementary system, for internal events.

b) Dependency among elementary systems: events or causes of failures that create dependency among unwanted events of numerous elementary systems; we can distinguish various types of these dependencies:

- Functional dependencies, resultant from the facility design;
- Dependencies connected to common equipment, for instance, energy feeds (electricity, hydraulic power, pneumatic energy);
- Physical interactions, when the failure of a component will affect all close components (explosions, fires, bullet emissions, dynamic effects of burst pipes)
- Dependencies due to human actions, as a result of man intervention in all phases of elementary systems lives (design, manufacturing, installation, operation, maintenance)

c) Dependencies among components: similar to the previous, however affecting components of the same elementary system.

1.2.3 Common Cause and Cascading Failures

Common cause failures are those dependent failures whose origin is the same direct cause. They are therefore a subset of dependent failures.

As "same cause", one can identify, among others:

a) Event connected to environment;
b) Human error during different phases of the life of components;
c) Human error; and
d) Another component failure.

Common mode failures are common cause failures that appear on the same failure mode of components. These failures represent a subset of common cause failures. Concerning the components which must be 'a priori' identical or very similar in order to develop the same failure mode.

Cascading failures of order n are dependent failures that can be classified into chronological order A_1, ... , A_i, ... , A_n, such as each failure A_i ($i \neq n$) be the direct cause of the following failure A_{i+1}.

When these failures affect n components, it is said cascading failures of order n. Common cause failures and cascading failures mutually exclude each other. Common cause and cascading failures relate to the concepts of primary, secondary and command failure, previously shown.

1.2.4 Classification of Common Cause Failures

Five classes of common cause failures can be identified according to their cause [5]:

a) Environment aggressions: events related to the external environment of the system or internal to the system however external to the elementary system in question:
- normal environment of internal or external sources to the system (dust, salinity, humidity, temperature, vibration, aggressive atmosphere, radiation);
- external natural environment (extreme weather, earthquakes, floods);
- accidental environment of internal source to the system (environmental conditions resulting from an accident, whipping pipes, projectiles, local flood, local fire, local explosion);
- accidental environment of internal source to the system (aircraft crashes, flood by breaking dams, explosion in nearby facilities, fire in nearby facilities, close road and rail accidents, sabotage);

b) design errors: human errors occurred during the design of components and elementary systems, that compromise the system mission:
  - component functional unsuitability (sizing errors):
  - functional relationships of an elementary system presenting potential common cause failures;
  - inadequate or harmful periodic tests;
  - system or component difficult to operate;
  - system or component difficult to maintain.

c) manufacturing errors: human errors made during the manufacture of components:
  - non-conformity to technical manufacturing specifications;
  - inadequate adopted technology;

d) installation errors: human errors committed during the electromechanical assembly of components, at the factory or in the field, and during the commissioning tests.

e) operating errors: human errors made during inspections and periodic tests, maintenance actions and systems normal, incidental or accidental operation, affecting components or elementary systems previously recognized as suitable for the operation.

1.3 Operational Safety of Systems

1.3.1 Concept

The Operational Safety of Industrial Systems (OS) [6] today is a true engineering discipline, applied within all the different life stages of an industrial system, from its conception to its decommissioning, going through the stages of development and operation. The concept of Operational Safety of Systems can be seen by the diagram in Image 1.3.1-A.

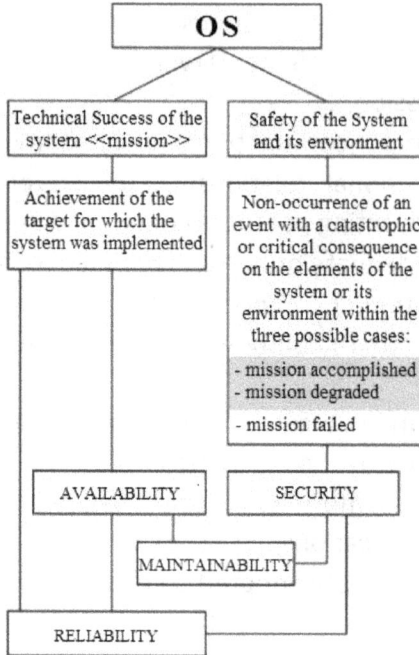

Image 1.3.1-A) Concept of OS

In a wider sense, the Operational Safety of the System can be defined as the "Science of Failures". It includes knowledge, assessment, prediction, measurement and control of a system failure.

In a strict sense, the Operational Safety of the System is the ability of a system successfully accomplishes the mission for which it was designed, without the occurrence of events with undesirable consequences for the components of the system itself and the environment where the system is in interaction.

The accomplishment of the "mission" means the performance of one or more required functions, within pre-established internal and external conditions to the system, and the consequent compliance with technical objectives for which the system was designed.

The non-occurrence of "events with undesirable consequences" should be guaranteed in the three possible cases of final results of the mission, which are:
   a) success (full compliance with the required functions);
   b) partial success (degradation without total loss of the required functions); and
   c) underachievement (total loss of the required functions).

"Undesirable consequences" are associated with the death or injury to the system operators (understood as those that can be individually identified), and with the public groups surrounding the physical implementations of the system, or indirectly affected by the system functioning (understood as those that cannot be individually identified), as well as unacceptable negative impacts (degradation of functions) to other systems that make up the environment in interaction with the industrial system under study.
The operational safety of the system can be characterized by the following component concepts (or techniques), that will be then individually defined:
   a) Reliability;
   b) Availability;
   c) Maintainability; and
   d) Security (or Safety per say).

Given these concepts, one defines the characteristics of operational safety of the system, shown by Image 1.3.1-B:
   a) MTTF: system meantime functioning before first failure ("Mean Time To Failure");
   b) MTTR: meantime to repair ("Mean Time To Repair");
   c) MUT: meantime functioning after repair ("Mean Up Time");
   d) MDT: meantime of unavailability, corresponding to the crash detection phases, the breakdown repair and return to service ("Mean Down Time"); and
   e) MTBF: meantime between two consecutive failures of a repairable system ("Mean Time Between Failures")

```
        FAILURE     REPAIR     FAILURE
           │           │           │
           ▼           ▼           ▼
┌──────────────────┬─────────┬──────────┐
│       MTTF       │   MDT   │   MUT    │ ──────► t
├◄───────────────►─┼◄──────►─┼─◄──────►─┤
│                  │                     │
│                  │        MTFB         │
│                  ├◄─────────────────►──┤
```

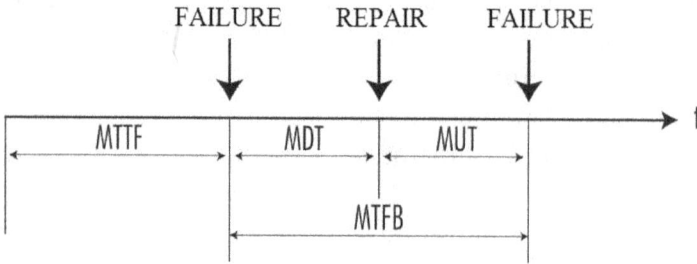

Chart 1.3.1-B) Characteristics of Operational Safety of a system

In general, other concepts have been introduced in specific studies, derived from four basic concepts, and can also be included in the Operational Safety of the System, such as:

a) Durability: the ability of an entity to remain in a state to fulfill a required function, within given conditions of use and maintenance, until a state limit is reached;

b) Continuity: ability of a function, once obtained, to continue to be guaranteed under certain conditions, for a wanted period of time;

c) Vulnerability: a system susceptibility to potential aggressions;

d) "Survivability": ability of a system to resist to internal and external aggressions.

The common point among the component concepts of Operational Safety of the Systems is the use of probabilistic language as a mean of assessment. However, when used in the field of Operational Safety of the System, this language loses in part its formal rigor of origin, established by the Probability Theory. Consequently, different quantitative assessments associated with a similar situation can be made by analysts of culture and diverse experience.

This fact is due to the implicit or explicit uncertainty that exists about the universe that serves to calculate the probabilities. The introduction of a conditional universe or Body of Keynes[7] made by different theorists of probabilities (among them Emile Borel[8]) associated with Decision Analysis on Uncertainty [9,10], leads to a subjective approach to probability, different from the formal mathematical approach [3,11].

1.3.2 Reliability

To sum up the numerous existing definitions, reliability is the ability of a system to fulfill a required function, within pre-established conditions for a certain period.

Reliability is usually measured by the probability of an entity E to perform a function required on imposed conditions during the period of time [0, t]:

$$R(t) = P (E \text{ without failure on } [0, t])$$

The opposing ability, which we shall call "Unreliability", can then be measured by:

$$\bar{R}(t) = 1 - R(t)$$

One must distinguish three types of measurement of different kinds which can be made of R (t):

a) Operational reliability (either observed or estimated), that results from observation and statistical analysis of the behavior of identical entities within defined operating conditions;

b) Extrapolated reliability, that results from an extension, based on extrapolation or interpolation hypothesis, of an entity operational reliability for periods of time and/or different conditions; and

c) Estimated reliability (or forecast), that estimates the future reliability of a system based on considerations about its design and about its components reliability.

Operational reliability is a statistic, extrapolated reliability is a model of a stochastic process and estimated reliability is a subjective probability, each have a specific use that should not be misinterpreted.

1.3.3 Availability

Availability is understood as the ability of an entity to be able to perform a required function within pre-established conditions and in a certain moment.

Availability is usually measured by the probability of an entity E to be able to perform a required function within pre-established conditions in a time t:
$$A (t) = P (E \text{ without fail at } t])$$

The opposing ability, which we shall call "Unavailability", can then be measured by:
$$\overline{A} (t) = 1 - A (t)$$

One must distinguish three types of measurement of different kinds which can be made of A (t):

a) instant availability (or immediate), which is the classic definition presented; in particular, this availability is the one that allows a system to face an emergency situation;

b) potential availability (or expected or continuous or statistical), which is the system ability to function continuously for a certain period of time, considering its current state; this availability is directly conditioned by the system maintenance and the inspection for its components; and

c) post-incidental availability (or survivability), which characterizes the ability of the system to continue to perform its functions, even in degraded or partial way, following an incident that has affected some of its parts.

1.3.4 Maintainability

Maintainability is understood as the ability of an entity to be maintained or restored in the state to perform a required function, when maintenance activities are carried out within pre-established conditions, following a set of procedures and previously prescribed means. The maintainability is usually measured by the probability that the maintenance of an entity E, performed in given conditions, with prescribed procedures and methods, is complete at a time t, knowing that the entity has failed at t = 0.

$$M (t) = P \text{ (maintenance of E complete at t)}$$

$$\text{or } M (t) = P \text{ (E is repaired on [0,t])}$$

The opposing ability, which we shall call "Non-maintainability", can then be measured by:

$$\overline{M} (t) = 1 - M (t)$$

This concept only relates, obviously, to repairable systems. The maintainability characterizes the system ability to restart and fulfill its functions after a failure.

At this point, it becomes very important to distinguish the three basic types of maintenance activities:

a) preventive maintenance: the one that is made according to a pre-established plan, based on the system functioning history, and seeks to prevent the occurrence of a failure;

b) predictive maintenance: the one that is made at the moment when a monitored variable (temperature, pressure, vibration, physical and chemical composition of process fluids, current, voltage) of the system reaches a critical value preset, and seeks to restore the normal functioning of the system before it degrades to an unacceptable level;

c) corrective maintenance (or repair): the one that is made after occurrence of a failure or unacceptable degradation of the system, and seeks to restore its normal functioning conditions.

1.3.5 Security (or Safety)

Security or Safety per say is understood as the ability of an entity to avoid the occurrence, within preset conditions, of critical events for its functioning or catastrophic for its operators and environment. The security is usually measured by the probability of an entity E not to give rise, within certain internal and external conditions, the occurrence of a preset series of catastrophic consequences, over its useful life.

$$S (T) = P (E \text{ without failure} \Rightarrow \{C\} \text{ at } [0,T])$$

The opposing ability, which we shall call "Insecurity", can then be measured by:

$$S (t) = 1 - S (T)$$

1.3.6 Cindinistic

Theoretical concepts and applications of the language of probabilities presented in this work are directly associated with its use in the concept of Security of Operational Safety of Systems, treated within the general discipline of Cindinistic or Science of Danger [12]. In fact, the probabilities used are several orders of magnitude lower than those used by the concepts of Reliability, Availability and Maintainability, other main components of Operational Safety of Systems that constitute a "adapted norm" to rare or improbable events and not to statistics. Consequently, even if certain mathematical concepts and techniques (probabilities and statistics) are common to the other components of Operational

Safety of Systems, the basic concepts are specific and the techniques are suited to the Systems Security.

1.3.7 Safety Levels

When the concept of Operational Safety of Systems is applied to the preliminary design of a facility is that events such as combined or not combined failures with human errors or external aggressions or threats that may lead to an unsafe state will be identified. After that, statistical and probabilistic methods will allow to evaluate the probabilities of the resulting failures and insecure states, through their spread and transformation within the system, in the form of a scenario.

Analyses of the Operational Safety of Systems will then be developed with three distinct and complementary objectives that can be associated with the security levels of a facility [13]:

a) Primary Level (Level 1): Fulfillment of technical objectives for which the system was designed (total or partial success of the mission); this level implicitly contains the "capacity" of the system, being an engineering activity to which the design techniques and employed technologies associated; it is associated with the level of prevention of the system regarding the occurrence of initializing events of an undesired event;

b) Secondary Level (Level 2): it is associated with the level of protection of the system regarding the occurrence of an undesired event; in other words, the system ability

c) Tertiary Level (Level 3): it associated with the gravity of an undesired event effects over the system and the environment, that is, how the consequences of the undesired event are felt within the system itself and in the environment, involving the application of possible internal and external emergency procedures to the system.

1.3.8 Commitment between Reliability and Safety

Reliability (and, in certain circumstances also availability) may conflict with security. This occurs in situations where it is necessary to prioritize between the possibility of success of the mission and the possibility of occurrence of a catastrophic event.

A very simple example may help visualize this type of situation: a firing system of a weapon, based on the closing of a relay. This relay has two types of failure:
   a) untimely closure, with a probability of 0.01; and
   b) open locking, with probability 0.02.

Considering that the untimely closure is a catastrophic event and that the open locking implies in the failure of the mission, one could adopt three types of solution:
   a) without redundancy: in this case, reliability would be 0.97 and security would be 0.99;
   b) redundancy parallel: in this case, reliability would be 0.9797 and security would be 0.9801;
   c) redundancy series: in this case, reliability would be 0.94 and security would be 0.9999.

Observe that the adoption of one or another type of redundancy implies privileging one of two types of breakdown: reliability requires redundancy parallel and security requires redundancy serial.

Another example would be a system of a submarine, whose proper functioning is necessary for the safety of the ship. One can imagine three possible structures derived from the operating procedures, summarized in Table 1.3.8-A, which considers, in all cases, that the reliability of the system throughout a mission is 0.999.

This example also shows that Reliability and Security may conflict, as relative demands of these two concepts may lead to different technical solutions, in some cases incompatible:

Table 1.3.8-B summarizes the differences and common points between the two concepts:

Single System simple failure ⟹ accident	Reliability success: 0.999 non-repair: 0.999 Security: 0.999
System with simple redundancy simple failure ⟹ underachivement double failure ⟹ accident	Reliability success: 0.998 non-repair: 0.998 Security: 0.999999
System with double redundancy simple failure ⟹ repair double failure ⟹ underachivement triple failure ⟹ accident	Reliability success: 0.999997 non-repair: 0.997 Security: 0.9999999

Table 1.3.8-A) Conflicts Between Reliability and Security

Clearly, people did not wait until the development of the Operational Safety of Systems discipline to analyze safety and enforce rules that would protect people, the environment and assets against excessive risk. In multiple domains there are long ago safety rules and standards. The naval industry stands out, as pioneer in the elaboration of these standards and rules, through the classifier societies of ships (Lloyd Register, American Bureau of Shipping, Germanischer Lloyd, Bureau Veritas, Det Norske veritas, among others).

Safety rules and norms generally involve the identification of a certain number of critical cases and the imposition, on a regulatory basis, of design norms that allow us for maintaining a suitable level of safety in such cases. Critical cases are defined from the analysis of real observed situations and that lead to (or could have led to) an accident. Therefore, design norms constitute technical solutions whose experience has shown, at least within a certain knowledge domain, adequate results.

DIFFERENCES	
RELIABILITY	SECURITY
1. It is applied, in general, to a component or item considered in isolation	1. It necessarily applies to a system and emphasizes interaction problems
2. Generally does not take human factors into account	2. It considers man as a part of the studied system
3. It studies the proper functioning of an entity under well specified operating conditions	3. Must address any circumstance, normal or abnormal, which could lead to dangerous situations
4. It involves failures whose cost is of the same order of magnitude as the normal cost of the mission	4. It involves events whose cost is of an order of magnitude clearly higher than of the normal cost of a mission
5. It involves events that can be generally assessed by experience and statistical treatment	5. It involves multiple combinations of events, individually unlikely, that escape a conventional statistical treatment
6. It may conflict with Security and require in certain cases different technical solutions	6. It can be improved by multiple means, with reliability increasing from certain parts of the system as just one of them
SIMILARITIES	
RELIABILITY	SECURITY
7. They use probability calculation and statistical methods	
8. They are established by comparable means (specifications application, quality program, project reviews, collection and treatment of technical information)	
9. Reliability and security analysis and programs span the entire life cycle of a system, from its design to its decommissioning	

Table 1.3.8-B) Comparison Between Reliability and Security 1.3.9 Classic Safety Standards

Even though abstracting from all regulatory aspects, such design norms are an extremely useful reference in the design and development phases, where we still cannot carry out more precise analysis. It is necessary, however, to know as soon as possible the extent to which the techniques used in the project move away from the domain of knowledge encompassed by the gained experience, and may therefore potentially derogate from the technical conditions implicitly or explicitly assumed by the safety regulations and norms.

If a system has an operating domain close to that of similar systems that preceded it, if it is designed and manufactured according to similar procedures and if it satisfies the same technical conditions, the same regulations, the same norms, then it may be considered as having equivalent security. So, safety norms offer, in a rational way, acceptable means of demonstrating safety for classic solutions.

Consider that safety regulations and norms have a legal scope: they serve as basis for the evaluation of systems that must obtain governmental authorization to operate (approval and licensing), as well as assist in establishing the responsibilities of the parties involved in case of an accident.

However, there are major inconvenient to dogmatically enforce safety regulations and norms, because the rapid evolution of technology and the diversity and specialization of systems make it very complex to develop "universal" application procedures.

a) cases considered "a priori" as critical are not always in reality the most critical; their identification is usually based on generally implicit assumptions about the use of the system and the reliability of its various parts; the very notion of a critical case defined "a priori" can be questioned, since its identification should be preceded by an analysis of the reliability of the system different parts;

b) certain norms that have proven to be satisfactory for systems from previous generations may become an obstacle to those of new generation, considering evolutions that have occurred in the operating procedures and at the technological level;

c) transforming a technical solution, a design procedure valid at a given historical moment, into an unchanging requirement leads to technical sclerosis, halts progress and constitutes in the long term an impediment to improving safety;

d) the mere statement of critical cases and design procedures, as generally found in regulations and norms, does not make it possible to discern the security objectives that were targeted; and

e) when a particular requirement is imposed on a part of a system, it is necessary, in order to interpret this requirement, to consider with great emphasis the integration of this part into the system and the use of the system.

1.3.10 Safety as Quality

Quality is the measure within which a particular product is adapted to requirements for which it is intended to be met [14]. Norm [15] defines it as the "ability of a product or service to meet the needs of users".

Strictly speaking, the quality of a product is characterized not only by its compliance to specifications which define it, but also by its ability to remain in compliance with its specifications throughout its useful life. One of the fundamental characteristics of a product that contributes to its quality is reliability, that is, its ability to retain its original characteristics.

However, a rather widespread approach considers quality as the product compliance to its specification on an ex-factory basis. Reliability would then be its ability to remain compliant over the period of use [16]. In this sense, reliability becomes an extension of quality over time.

In reality, the concept of Operational Safety of Systems is closely connected to the concept of Quality, more precisely, to Quality Assurance. Reliability, availability, maintainability and security are ultimately the means to ensure that a "product" (understood here in its broadest sense) satisfies, throughout its useful life, the needs of its users, which necessarily include non-occurrence of damage to users themselves and to the environment within which such use is made. It should be emphasized, however, that the Quality Assurance led to specific theories and methods [17] that will not be addressed by the present work.

1.3.11 Safety and Decision-Making Process

Decision is understood as any choice, among many possible actions that carry failure risks of the "mission" and/or damage to people, property and the environment. In these terms, the decision-making process can be interpreted as risk management [10].

All safety analysis and evaluation techniques of systems have the same goal: to provide the decision maker with elements that allow him to make the best possible decision, based on the set of information known and considered in the safety analysis and summarized by the summary of results presented to it.

The determination of the "best possible decision" or "great" decision is made by the implicit or explicit application of a cost-benefit analysis procedure in which risks constitute a cost to be considered or, conversely, risk reduction is a benefit.

A theoretical basis for elaborating a rational decision-making procedure can be derived from the general decision theory, or probability-utility theory that results from the confrontation of statistical decision functions [18], and from neo-Bayesian theories [19]. Within this context, decision-making statistic groups a set of methods whose target is rational decision-making in the presence of uncertainty [20].

1.3.12 Murphy's Laws

The ability to appreciate an undesirable event and the perception of the risks involved is intimately associated with the experience gained by the analyst and the "decision maker". Captain Murphy of the US Army [6] has expressed, in a funny way, a list of fatalistic axioms that describe in a deterministic and negative way diverse behaviors and situations. Twelve of these axioms are presented below:

- Anything that can go wrong, will go wrong;
- If more than one thing can go wrong, the one that will go will be the one that is most catastrophic;
- A hidden flaw will become more apparent in the worst circumstances;
- A shortcut to perform a dangerous operation is the fastest path that leads to disaster.
- Any task that can be done in an incorrect way, no matter how reduced the probability is, one day will be made this way.
- Every piece that is susceptible to a failure will fail at the most inopportune and harmful moment.
- No matter how difficult it is to damage an equipment: a way will be found to do so.
- Predicting the worst is generally the best thing to do.
- An infinite number of people will appear from an infinite number of places within an infinitesimal period of time to say what should have been done to prevent the accident before it has occurred.
- Nothing stupid proof can be done: stupid people are very ingenious.
- After tightening all thirty-six bolts of a flange, it will be noticed that the gasket has been forgotten.
- Two identical units, with identical functioning, under identical test conditions, will function completely different after they are installed.

These "axioms" very accurately express the fact that, in safety, if a situation is "a priori" possible, it is necessary to consider it probable in the analysis. Consequently, one must try to limit their probability of occurrence or to minimize the seriousness of their consequences, and not try to ignore it at the cost of a great effort of justification whose verification is almost never possible.

1.4 Risk

1.4.1 Concept of Danger

Traditionally, danger is defined as a potential damage that can fall on people, assets or the environment. This notion is general, and must correspond to a situation sufficiently well determined and described in a degree of detail adapted to the intended use of this concept.

This definition may refer to dangerous situations of random nature, of natural, technological or economic origin, or to threatening situations of a deterministic nature, linked to intentional actions or acts of sabotage.

The concept of danger is based only on the potential consequences of an undesirable or feared event E, without considering the real possibilities that it will actually occur. These consequences can in turn be classified in a hierarchical scale of severity g: a danger will be greater as the consequences will be more serious. Danger is then a one-dimensional concept, associated only with the severity of the consequences of event E.

$$P = P\ (g)$$

1.4.2 Concept of Risk

Risk is the perception or evaluation of the possibilities of the actual occurrence of an undesirable event E that leads to the realization of a danger, which by definition is something potential, that is, that has not yet occurred. As has been seen, a danger is associated with the occurrence of consequences with a certain severity.

Unlike danger, the concept of risk is then based not only on the consequences (and their severity), but also on the occurring possibility of these consequences. Specifically, risk is a two-dimensional concept more encompassing than danger, which combines both the possibility and the severity of an event E.

$$R = R \, (p, g)$$

This two-dimensional character of risk makes its hierarchy impossible: set theory shows that there is no relation of order in R^2, a set of conjugates of real numbers [21], which concludes that, in theory, (p_1, g_1) and (p_2, g_2) cannot be compared.

1.4.3 Risk Quantification

In order for the concepts of danger and risk to be effectively used in Operational Safety Systems, it becomes necessary to quantify p and g. By establishing a functional relation that conjugates them for the formation of R, one can get around the theoretical problem of hierarchy.

In order to quantify the possibility p, the most used method is to associate a probability $P \, (E)$ with the occurrence of event E. The quantification of the severity scale, which makes it possible to measure g will depend on the nature of event E, because as it was seen, danger affects people, assets and the environment. For events whose consequences are of the same nature, the establishment of this scale becomes relatively simple, for example:

a) for a danger whose consequences correspond to a single type of injury (such as death), affecting only a homogeneous group of human beings for which somatic responses to similar injury are expected (such as the group of facility operators), the obvious scale is generated by the number of people affected;

b) for a danger whose consequences correspond to a single type of injury (such as crude oil spill into the environment), affecting only a homogeneous and defined area (such as a certain and limited coast), the obvious scale is generated by the quantity released into the environment;

c) for a danger which affects exclusively material assets and properties, the obvious scale is the monetary scale of costs associated with losses and damages.

These ideal conditions, however, very rarely approach the reality of a security analysis, since among other aspects:

a) dangers usually affect people, assets and the environment at the same time;

b) groups of people affected by danger are not homogeneous (the so called "public", which is not individually identifiable, involves several age groups, both genders, eventually debilitated or particularly susceptible people, implying different sensitivities to the effects of the events incurred);

c) injuries to people are of various types, whose relative severity is difficult to compare, such as:

- temporary and permanent psychological disorder;
- temporary and permanent invalidity;
- immediate death e or in the long term after the event;
- delayed effects, of a generally stochastic nature, occurring in the long term after the event, such as:
  - increased possibility of contracting cancer due to the incorporation of chemicals and/or exposure to ionizing radiation;
  - increased probability of genetic disorders occurring over future generations.

d) releases to the environment are made of more than one product, with different hazards on the public and other forms of animal and plant life;

e) physical areas of the environment that receive the releases are not homogeneous, that is, they have different ecological responses to aggressions, nor are they perfectly defined

For these reasons, the scales of consequences used in reality have a strong subjective connotation and generally do not cover all the consequences associated with an event, retaining in general a single type of consequence, considered the most serious.

For relatively minor consequences, which do not involve immediate death of identifiable people or irreparable damage to the environment, such as those involving delayed stochastic effects, with exposures to small doses of ionizing radiation, there is a tendency to adopt scales of monetary severity [22,23].

It is important to point out that the establishment and quantification of gravity scales of accidental catastrophic events are a problem that goes beyond engineering techniques, referring also to economics, epidemiology, sociology, ecology, among other social and biomedical sciences. These scales raise several ethical and deontological questions, which engineering cannot fail to consider.

The third important aspect in order to apply the concept of risk in the Operational Safety of Systems is the establishment of a functional relationship that conjugates p and g, assuming them numerically determined, in a unique measure for the formation of the value of R, and its consequent one-dimensional quantification.

$$R = f\,(p,\,g)$$

A rational function considered for this relation is the mathematical expectance of the consequences, that is, the product (p $_x$ g). For example, if there is a 0.001 probability of an event occurring resulting in the loss of 1,000US$, there is a risk of 1US$, implying that another event of probability 0.1 and loss of 10US$ would have the same risk.

In this model, it is assumed that risk is independent of p and g, depending only on its product (p $_x$ g), that is, for a constant probability, the risk increases linearly with severity and vice versa, or yet in a chart (p, g), the iso-risk curves are straight, as shown in

Image 1.4.3-A.

For relatively high probabilities, for classic concept or concept of relative frequency and, even more important, for consequences with limited gravity, the use of this function presents no special problems, besides those associated with the rigor of the modeling of the system being studied.

However when events are rare, in other words, very reduced and conceptually inductive (probability as verisimilitude) probabilities with severe consequences, including, for example, death of a large number of people, there are important restrictions on their use:

a) A third risk component, mentioned up to now, but not considered, becomes preponderant: individual or collective perception of the dangerous situation and the event E that causes it; and

b) Ethical and deontological considerations that give rise to the expectance model, due basically to the application of a technocratic rationality to problems of a strongly social and psychological, and ultimately political, character; these considerations go completely beyond the scope of engineering itself, but the technicians involved in the safety analysis of systems that can provide this type of risk must be aware of their existence, knowing that decisions taken at the individual or collective level in the area of prevention of events with severe consequences escape technical rationality.

Individual or collective risk perception may, in theory, be integrated into a function that combines probability and severity. The mathematical expectance model of consequences assumes that the perception is independent of severity. Then it is possible, through sociological research methods [24], to model the aversion to severe consequences, which is empirically observed in people and in societies. Notice that individual aversion can differ significantly from the collective: for example, the same person, individually, can be a smoker and collectively reject the installation of a factory that is supposed to be polluting in her neighborhood.

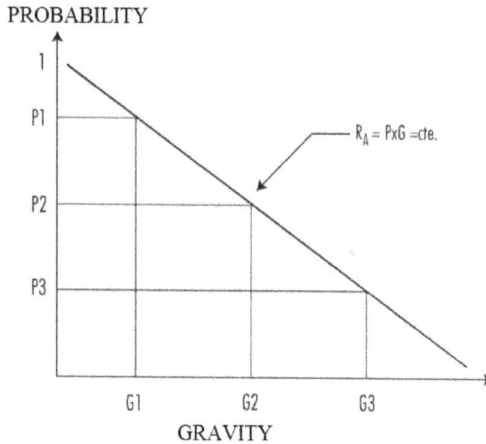

PROBABILITY

$R_A = P \times G = cte.$

G1   G2   G3

GRAVITY

Chart 1.4.3-A) Iso-risk Curves

One of the techniques used to model aversion to severe consequences is the so-called "consent to pay" [25]. A model of this type would distort the iso-risk curves making them approximately asymptotic with regard to the maximum plausible consequence, according to Image 1.4.3-B.

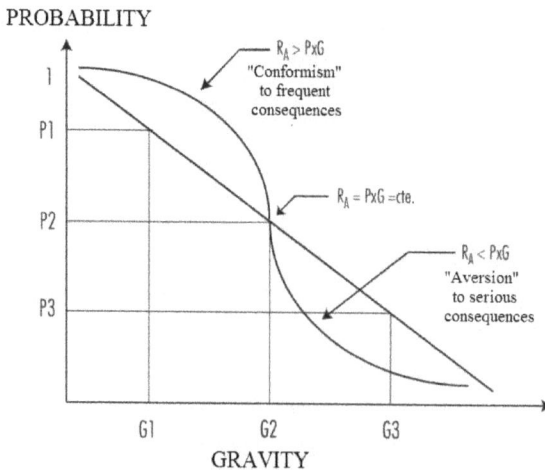

PROBABILITY

$R_A > P \times G$
"Conformism" to frequent consequences

$R_A = P \times G = cte.$

$R_A < P \times G$
"Aversion" to serious consequences

G1   G2   G3

GRAVITY

Chart 1.4.3-B) Distortion of Iso-risk Curves due to Social Perception

Chart 1.4.3-C presents the results of a research on risk perception [47].

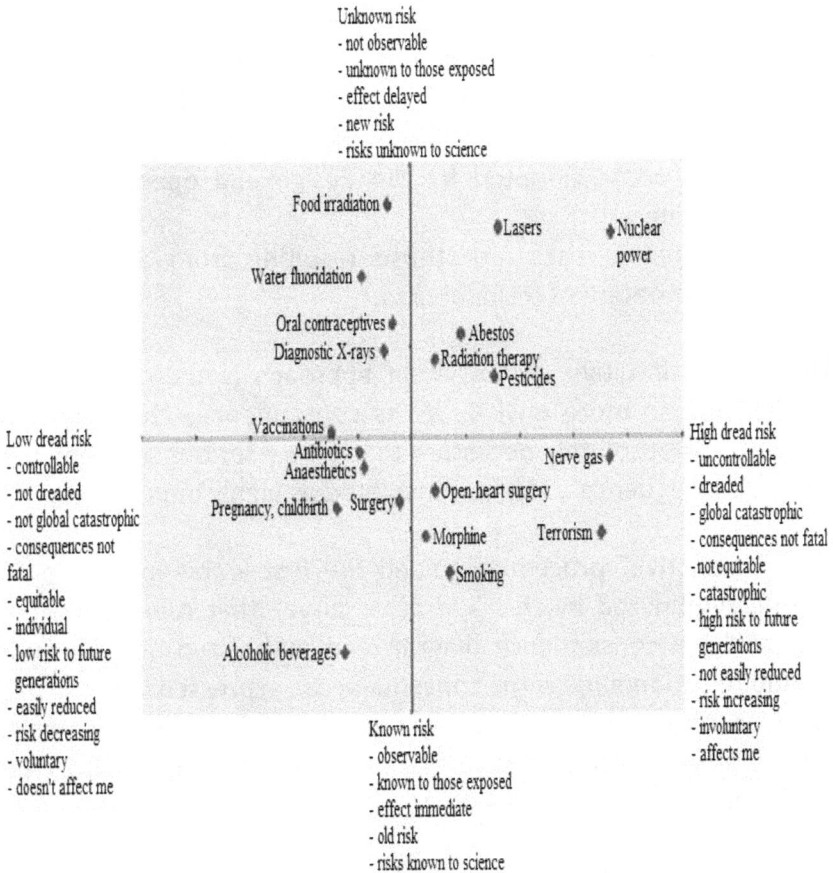

Unknown risk
- not observable
- unknown to those exposed
- effect delayed
- new risk
- risks unknown to science

Food irradiation ◆

◆ Lasers   ◆ Nuclear power

Water fluoridation ◆

Oral contraceptives ◆
Diagnostic X-rays ◆

◆ Abestos
◆ Radiation therapy
◆ Pesticides

Low dread risk	Vaccinations ◆	High dread risk
- controllable	Antibiotics ◆	Nerve gas ◆	- uncontrollable
- not dreaded	Anaesthetics ◆	- dreaded

◆ Open-heart surgery

Low dread risk
- controllable
- not dreaded
- not global catastrophic
- consequences not fatal
- equitable
- individual
- low risk to future generations
- easily reduced
- risk decreasing
- voluntary
- doesn't affect me

Vaccinations ◆
Antibiotics ◆
Anaesthetics ◆
Pregnancy, childbirth ◆ Surgery ◆
◆ Morphine
◆ Smoking

Nerve gas ◆
Terrorism ◆

Alcoholic beverages ◆

High dread risk
- uncontrollable
- dreaded
- global catastrophic
- consequences not fatal
- not equitable
- catastrophic
- high risk to future generations
- not easily reduced
- risk increasing
- involuntary
- affects me

Known risk
- observable
- known to those exposed
- effect immediate
- old risk
- risks known to science

Chart 1.4.3-C) Social Perception of Risk

1.4.4 Risk Classification

Considering that all human activities, notably complex industrial activities such as the operation of nuclear facilities and ships, are associated with many kinds of risks, related to different harmful effects to individuals, society, environment, material assets and properties, it is possible to deduce that zero risk is a totally unrealistic abstraction.

The risk assessment associated with the operation of an industrial facility leads us to distinguish three types of risks:

a) potential risks: are those to be feared in the absence of all prevention and protection measures, being associated with the previously presented notion of danger;

b) residual risks: are those that subsist despite the accident prevention measures and mitigation of consequences, if they occur, adopted by the design and operation of the system;

c) acceptable risks: are those resulting from a process of optimization of residual risks.

The logical, intuitive, optimization approach is then to make an accident be so more unlikely as its consequences could be more serious. It is absolutely necessary that a very serious accident, with severe consequences, can be considered as highly improbable.

This "instinctive" procedure guided the first works in the field of safety, symbolized by the "Farmer's curve" that represents, on a probability x consequence diagram, an authorized domain and a prohibited domain, with consequences expressed in assumed values for the tailings of radioactive material made by a nuclear facility after an accident (the curve of Farmer's original work [26] dealt with Iodine-129 tailings).

The designers of nuclear power plants then sought to deepen and specify this curve, determining the pairs (probability; radiological consequences) considered as acceptable. Afterwards, safety organisms set indicative limits of the maximum probability of accidents susceptible to cause consequences considered to be unacceptable, with even lower probability situations considered as the inevitable residual risk.

Once these targets were established, it remains to demonstrate that all types of accidents judged to be plausible were considered and analyzed by the accident studies of the facility and that the safety and emergency systems of which the facility is equipped effectively allow to meet this target.

1.5 Absolute Safety and Acceptable Risk

1.5.1 Absolute Safety

The absolute (or total) safety of an activity or system corresponds to the impossibility of a catastrophic accident at any given time (at present or in the future) and regardless of the state of the system or its environment for the set of all failures, human errors and possible external aggressions.

Such a project would then require the "perfect and exhaustive" knowledge of the system, as well as quality problems in the design, fabrication, construction and operation, and of all its possible states, as well as the consideration of all outside environments, even the most extreme and rare ones.

This working hypothesis is obviously not reasonable, either for reasons of scientific knowledge or for technological reasons and, more simply, for reasons of human imagination. This translates into one of the basic principles of system safety:

"Absolute safety is a utopia."

This primary notion of "absolute and right safety", which aprioristically corresponds to the impossibility of catastrophic accidents, must be replaced by the notion of "objective safety", concerning an acceptable risk with regard to realistic financial and technical efforts expended.

1.5.2 Risk Acceptability

Although the approach described above is very clear and objective, it presents difficulties to be put into practice because it is based on the concepts of "plausibility" and "acceptability" that are directly related to the individual and social perception of risk.

First of all, the definition of "plausible accidents" can be extremely controversial. For example, is the fall of a small civil airplane over an industrial facility plausible? What about a Jumbo 747? What about a military fighter aircraft? What about a meteor? The answer to these questions depends on the level of socially acceptable residual risk, which in turn depends on where the facility is located and the statistical significance of the historical database on similar events occurred on the place.

On the other hand, finding a social consensus about the acceptable level of residual risk is one of the most difficult tasks because the psychosocial perception of risk varies appreciably from person to person and from community to community.

When the risk is evaluated as the product probability x consequence, that is to say, according to the mathematical expectance model of consequences, and is thus used to define limits of acceptability, infinite individual questions can be generated, because at the individual psychological level, the risk will hardly be perceived in the same way, which explains the eternal controversy over the safety of nuclear facilities around the world (apart from certain political aspects that we are not here to discuss). The question "how safe and sufficiently safe" always remains without definitive social response.

Another important point still remains: in addition to the consequences of accidents at the individual and collective level, a serious accident involves other consequences, such as pollution of the environment, restrictions on the use of land and its products, social problems related to population evacuation, among others. Adding all these consequences to a unified risk assessment is one of the most complex or even impossible tasks.

Finally, the practical assessment of the probability of highly improbable events is extremely imprecise, which removes all character of absolute value. Its usefulness is restricted to the comparison with values, calculated by the same methodology, obtained for other similar installations.

1.5.3 Risk Tolerability

Individuals tolerate different levels of risk depending on the benefits they image to gain by taking them. Likewise, the social tolerance of different risks varies significantly due to several reasons, some objective and liable to scientific evaluation and others totally subjective, derived from complex psychological attitudes.

Frank Layfield introduced the term tolerability in his report on the public hearing of the Sizewell B Nuclear Power Plant in Great Britain [27]. Tolerable does not mean acceptable: it implies the maximum limits of acceptability, while acceptable implies a voluntary acquiescence. Tolerating a risk does not mean that it is negligible, or that it can be ignored, but that it is necessary to keep it under constant review and reduce it even further, if technically and economically possible. There is also a clear implication that some risks simply cannot be tolerated.

A consolidation of the risk tolerability philosophy was developed in Great Britain, in response to the considerations made at the Sizewell B public hearing [28]. One of the major difficulties of this philosophy lies in the fact that safety regulations are based on the need for risks to be reduced to a level that is as low as reasonably practicable - the principle of ALARA ("as low as reasonable achievable") optimization [22] . Theoretically, this implies a continuous reduction of risk, in which the cost of this reduction, no matter the degree of risk involved, is compared with the benefits obtained, in terms of reduction of consequences, or "detriment", according to proper jargon). When costs become disproportionate to benefits, the system can be considered to be at an acceptable level.

However, it must be recognized that there is always a maximum level of risk that simply cannot be tolerated, except in extraordinary circumstances. The ALARA Principle can only be applied when this limit is implicitly or explicitly set. Likewise, there is a minimum level of risk for a system that can be considered negligible when compared to the risks associated with other human activities.

The ALARA principle will consequently be applied between the upper tolerance limit of the risk and the lower limit where this risk can be neglected. As the risk is reduced, the activity (or system) becomes progressively more acceptable and, consequently, less effort must be expended to reduce it. These concepts are represented by Image 1.5.3-A.

In order for such a model to be effectively used in Systems Safety, it is initially necessary to quantitatively define the levels of tolerability and negligibility. After that, a methodology must be developed to demonstrate that the safety of a system conforms to this model.

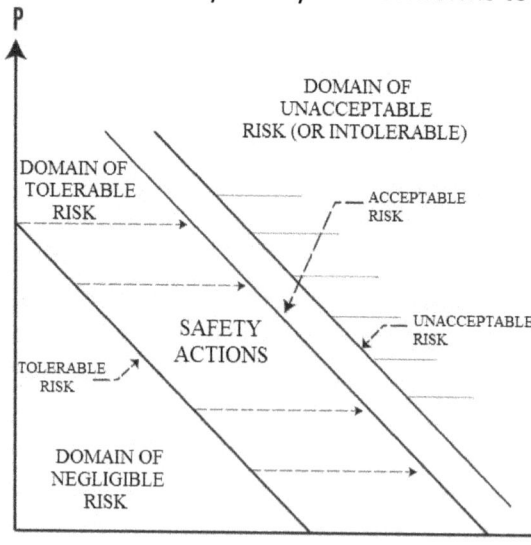

Image 1.5.3-A) Model of Social Tolerability to Risks

1.5.4 Commitment Between Local Risks and Global Risks

As has been seen, the risks affect basically people, the environment and assets or properties. With regard to people, one can, however, discriminate at least three human groups that can be affected, in increasing order of severity:

a) the operators and workers inside the dangerous facility;

b) the populations most directly affected by the potential consequences of the industrial facility operation, generally living near it ("critical group"), which in some cases benefits from the existence of the facility, in terms of jobs and social infrastructure; and

c) the general public, affected in an indirect and generally not very severe manner, by the potential consequences of the facility operation.

The risk perception and consequently its acceptability varies appreciably for each group. The power of political pressure from each of these groups will significantly influence the technical options adopted for the facility. Care should therefore also be taken to verify that a particular safety option that reduces the risk of one of the groups does not disproportionately increase the risk of others.

Regarding the environment, it is also possible to identify physical areas of incidence of the consequences of the industrial facility operation: local, regional, macro-regional and global. The risk perception also varies appreciably according to the extent of the potentially affected area. For example, an individual (or social group) may be strongly motivated to advocate actions that are supposed to reduce the so-called global "greenhouse effect" and be relatively indifferent to urban sewage pollution from the river that flows past their home.

Likewise risks to people, the power of political pressure from groups defending the environment in the different affected areas will significantly influence the technical options to be adopted for the facility. Care should therefore also be taken to verify that a given safety option that reduces the risk to one area does not increase, directly or indirectly, disproportionately the risk to others. These two aspects are highly significant for the determination of the management strategy of normal and accidental tailings and effluents from industrial facility.

1.5.5 Economic and Financial Aspects

The third area of incidence of risks of an industrial facility refers to individual and collective assets and properties, involving economic and financial aspects [29].

The occurrence of an undesired event in an industrial facility implies significant material losses:

a) losses of production due to the interruption of the operation, which may be more or less time-consuming, depending on the event and the type of facility, even reaching the final decommissioning;

b) losses of capital due to the damages caused by the undesired event and expenses associated with the repair of these damages;

c) losses of real state value due to the possible devaluation of the shares of the company to which the damaged facility belongs;

d) losses due to indemnities and aid to be paid to people directly and indirectly affected by the accident, whether they are workers of the facility or the general public; and

e) losses due to possible actions of evacuation, decontamination and recovery of internal and external areas to the facility.

Analyzing the problem under these aspects, it is noticed that safety is an investment that propitiates significant paybacks, that is to say, the non-occurrence of these losses. However, investment resources are by definition limited, and therefore safety also has a limit, especially when it is observed that the purpose of an industrial facility is production at competitive prices, not just safety.

It should also be noted that every productive process implies external costs [30], not included in production prices, which are indirectly borne by the society as a whole. The risks of the manufacturing industrial facility constitute one of these externalities. A valid sociopolitical strategy for improving the safety of industrial facilities would be the internalization of the risks cost associated to the safety. Such a policy would benefit, in terms of prices, the facilities with the best safety standards.

2. Probability in Systems Safety

Probabilities are one of the means of measuring a dangerous situation which may lead to the occurrence of one or more undesirable events. However, the concept and use of probabilities in security may seem very distant from probability theory as defined axiomatically [3,11].

2.1 Use of Probability

2.1.1 Probability Theory

From a measurable set Ω of elementary events [31, 32], called set of eventualities, of the possible or fundamental, we form the set of parts of Ω, which is noted as **A**, using the union and the complementation as laws of internal composition, for which **A** must be stable (closed) for all enumerable unions of events.

From the point of view of events, the set **A** comprehends:

- Ω, certain event;

- $\emptyset$, impossible event; and

- all random events formed from the elementary events of Ω and from the laws of internal composition.

A probability is an application of **A** over the interval [0; 1], which satisfies the three following axioms:

a) $1 \geq P(A) \geq 0$ for all $A \in A$ and $P(\Omega) = 1$

b) for every finite family $\{A_i, i \in I\}$ of disjoint events 2 to 2, we have:

$$P(\cup_i A_i) = \sum P(A_i) \text{ (additivity)}$$

c) for every sequence $\{A_n, n \geq 1\}$ of events descending towards

$\emptyset$, that is, $A_1 > A_2 > A_3 ...$, and that $\cap_I A_n = \emptyset$, we have: $\lim_{n\to\infty} P(A_i) = 0$ (sequential monotonous continuity in $\emptyset$)

A random variable is then an application of Ω in $\Re$, set of real numbers.

2.1.2 General Aspects

The formal assessment of a probability on **A** initially requires perfect knowledge of the set of possible Ω and the complete identification of the events of **A**.

The search for this knowledge is simplified when Ω is defined by a finite number of values (for example, the number of faces of a dice). It can be limited to an overall description of the set of possible, as is the case of a test, which can lead to an indefinite number of values (for example, the height of a wave).

In systems safety, the set Ω corresponds to the integrality of functioning states and breakdown states (non-functioning and malfunction) of the system, to which its consequences are associated in all the environments for which its mission is (or will be) carried out.

For complex systems or poorly or incompletely known environments, the exhaustive knowledge of Ω_s is impracticable. The same can be said, in particular, for the set of combinations of internal (or external) causes of failures leading to serious damage. However, an "a priori" partition of A_s into states class of the system **S** during or after functioning can be performed according to a severity scale of consequences, as previously defined.

2.1.3 Knowledge Domain and Zone of Certainty

It is possible to define two domains related to the functioning of a system:

a) <u>Knowledge Domain</u>, within which it is possible to accurately describe all functioning and breakdown states and their consequences on the environment within which it evolves; the same applies to the loading and impositions of the environment on the system;

b) <u>Lack of Knowledge Domain</u> (or Ignorance) about the functioning states of the system or its environment.

If within the Knowledge Domain, it is possible to expect to evaluate with more or less precision the probability of a failure mode of **S** or one of its consequences, it is important not pretend to estimate the probability of a non or poorly qualitatively defined event . In other words, it is absurd to *probabilize* the unknown.

The Knowledge Domain can be structured in two complementary zones, visualized by Image 2.1.3-A:

a) Zone of Uncertainty; b) Zone of Certainty.

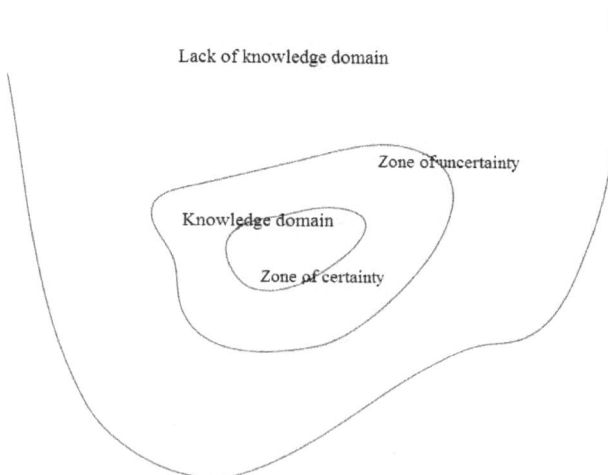

Lack of knowledge domain

Zone of uncertainty

Knowledge domain

Zone of certainty

Image 2.1.3-A) Domains of System Operation

The Zone of Uncertainty is that which corresponds to a qualitative knowledge of the system states associated with a random knowledge of each of them for a given situation. It is, for example, the state of a vehicle (perfectly identified) after 50,000km and, more precisely, its probability of reaching, "a priori", 50,000km. The uncertainty may be associated with imprecision over one of a system states, defined by magnitudes, associating a probability density with it. Its knowledge allows us to characterize deviations around an average. This approach is one of the bases of imprecise logic ("fuzzy logic") [33].

The Zone of Certainty corresponds to a deterministic knowledge of all the states of the system and its consequences. In general, this knowledge can be of two types:

a) <u>Theoretical Knowledge</u>, which is expressed through infinitely reproducible physical laws and many times validated by experience (Newton's laws, for example); this type of knowledge can be called theoretical determinism; note that this determinism can be expressed in the form of probabilistic laws, as in quantum physics;

b) <u>Statistical Knowledge</u>, which is expressed by statistical values deducted from observations, such as the average value of waves height in a given maritime region; this statistical determinism is only verified in an average of a large number of observations and cannot be applied individually to each observed element.

Technical decisions in practice involve the two areas of Knowledge Domain: for example, the timing of launching a satellite will be indirectly defined by the choice of launch site, which will have favored a meteorologically favorable site (Zone of Statistical Uncertainty), and directly by the immediate weather conditions on the launch site (Zone of Uncertainty).

2.1.4 Principle of Practical Certainty

Decision making in the Domain of Uncertainty is done by applying the Principle of Practical Certainty [34], regarding the consideration of almost impossible events. It is stated as follows:

"If the probability of any event E in a given experiment is small enough, it can be almost certain that when this experiment is performed only once, event E will not occur."

This principle adapts in the same way to the almost unitary probabilities, corresponding to right events.

It is evident that this "principle" cannot be mathematically demonstrated, but it is confirmed by everyday experience, which formalizes personal (subjective) experience to concepts of right and impossible. It is from this principle that the almost totality of the daily life decisions are taken. It causes zero-probability events to be ignored, considered "a priori" as impossible (shock of a large meteor with Earth) or probability near to one, considered "a priori" as right (dawn following night).

However it is important to emphasized that the almost certainty or almost impossibility of an event simply based on experience should be treated with extreme caution in the context of safety analyzes. In fact, the actual duration or practice of observation may be several orders of magnitude lower than which would be required to observe the undesirable event considered within the safety targets. Therefore, this event should not be "a priori" excluded from the analysis because it is only considered impossible the justification that it has never been observed.

For example, the observation of 50 years of accident-free operation of an industrial facility cannot reasonably meet a severe accident target of 10^{-7}/year, because the return period of the latter is 10,000,000 years. The probability of observing such an accident during the 50 years is $5x10^{-6}$, which makes it highly improbable. The same can be said of natural risks, where the implantation of residences in places of risk in general only takes into account the observations of the current generation, although historical reports indicate occurrences of serious accidental phenomena, such as floods, earthquakes.

In short, it may be emphasized that the proximity of an event is not reflected by its probability of occurrence. The principle of practical certainty is a corollary of the law of large numbers [3], which consists in stating that:

"During an experiment, when the number of tests increases indefinitely, the observed frequency of a given event considered as a possible result tends to a limit that is equal to its probability."

Within the principle of practical certainty, this probability is very close to 0 or 1. It is the great repetition of observations of a given event that leads to experience and the subsequent confidence that results from decision-making. The principle of maximization of entropy [35, 36] allows to draw from a set of observations related to the functioning of a system all the information it contains, in order to guide associated decisions.

Various regulations and procedures aim to eliminate all randomness in the design, implementation and operation of a system. The respect for these regulations and procedures allows associating options and actions with almost deterministic consequences that belong to the Zone of Certainty, allowing the application of the principle of practical certainty. One of the instruments of this action is the Quality Assurance.

For example, the design and construction of a pressure vessel from an established code (ASME, for example) are dictated by the need to eliminate elements of which one of the characteristics belongs to extremes of the probability distribution of a poorly known parameter, and therefore with uncertain value.

By force of habit of repeated successes, the decrease in vigilance may lead one of the functioning states of the system to the uncertainty zone, that is to say, where the causes of failures are of random nature, and therefore poorly controlled. It is well understood then that the safety of a complex system or more simply its proper functioning cannot be guaranteed within a certainty zone whose limits have been only established by the feedback, and therefore of the mistakes made and corrected of the past (but, obviously, also of successes).

To a small probability it is possible to associate the concept of rare event, defined as [37]:

"A rare event, in the sense of system safety, is an event that can lead to serious consequences to which the decisions that are made should allow you to signal a very small probability."

2.1.5 Notion of Chance

One concept that has been avoided to this point has been the one of chance, which classically sums up our knowledge. A literary definition, however quite significant of chance, is given by the Brazilian writer Millôr Fernandes:

"Chance is it the excuse our vanity gives to our ignorance."

Ignorance, and therefore chance, can be of two natures:
   a) lack of knowledge about the very nature of danger; and

   b) lack of knowledge about the process and the instant of realizing a risk scenario, although knowing the source from where the danger occurs.

It can then be concluded that the domain covering the concept of chance groups the Lack of Knowledge Domain and the Zone Uncertainty of Knowledge Domain.

From this fact, it is well understood that the introduction of probabilistic language into systems safety is an attempt to control uncertainty through a hierarchical measure of the occurrence of accident scenarios (causes and effects), this measurement can only be done on perfectly described and therefore known elements. This, in fact, excludes the "ignorance" side of chance, and hence its control. Thus, the elimination of the word chance in system safety is justified.

2.2 Different Definitions of Probability

There are essentially four ways of defining probability [6, 11]:

2.2.1 Classic Definition

Classic definition corresponds to the more ordinary definition we might call "instinctive": if in an experiment we identify **N** possible alternatives of realization and **N$_A$** alternatives that interest the experienced, then

$$P(A) = N_A / N$$

2.2.2 Axiomatic Definition (or Countable Measure)

Axiomatic definition is the one based on Kolmogoroff's Theory of Measurement, which only uses abstract mathematical concepts, not associating probability with any property of physical systems. This is the formal definition, presented by Probability Theory.

An application of the axiomatic definition is the countable measure, made by counting particular objectives contained in a set of observed objects and the ratio between the sizes of the two sets. The proportions thus calculated can be considered "a priori" as probabilities, since they verify the basic axioms. Such a count can be made on a manufacturing line, with the number k of defective elements being observed on a series of n. And the ratio k / n allowing us to calculate the defective elements in the manufacturing of the considered series.

A part of statistics called descriptive statistics, or exploratory statistics, is usually associated with this type of measurement. This last terminology corresponds to the fact that it serves as basis for the knowledge of the functioning of a system. Within this perspective, the countable measure is one of the elements of the feedback (Pareto's diagram, for example).

2.2.3 Relative Frequency

The definition of Relative Frequency is the most used one in engineering and physics, having been formalized by Von Mises: if an experiment is repeated **n** times, then the probability P(A) is defined as the limit of the relative frequency n_A / n when n tends to infinity.

$$P(A) = \lim_{n \to \infty} (n_A / n)$$

Frequency is a direct count (integer). Care must be taken to precisely define the fundamental set. Thus, the observation of three events per year cannot be directly translated by a probability, keeping year as time scale. On the other hand, a frequency of three observations in 365 days can be considered as a countable measure, and therefore a probability, if observations are considered on a daily basis, however being careful to previously verify that the duration of the event is "a priori "less than a day.
Considering the previous example on manufacturing control, the k / n ratio can only be taken as an estimator of the proportion of defective elements if the manufacturing conditions do not evolve over time.

The search for a probability from a frequency is another part of statistics called inferential statistics, whose objective is deduction, from a sample of the properties of the population from which it comes.

Consequently, it covers the determination of the parameters of laws of probability from observations samples, as well as the confidence intervals associated with them related to a statistical risk limit defined "a priori".

2.2.4 Likelihood

The definition of probability as likelihood, also called subjective probability, is a measure of confidence that can be given to an uncertain proposition. This definition has no relation to previous ones, since it is not deduced from direct observations that allow an effective measure.

The development of theory of decision in uncertainty gave rise to the concept of probability as likelihood, applicable to facts that may be outside the real possibilities of experimentation (for example, 10^{-7}/year) or even simple observation due to spatial or temporal impossibility.

Using an example from Morlat [37], historians may attach a weight to the statement "Julius Caesar was in England." The weight of this affirmation, normalized between 0 and 1, corresponds to the likelihood that each historian can estimate, from his own erudition and reflection: it is therefore a measure of the possible.
It should be noted that this weight is usually given with respect to complementary propositions, and not with respect to the set of possible. In the previous example, the set of possible is formed by two events:

$$E_1 = \{Julius\ Caesar\ was...\};\ and$$
$$E_2 = \{\ Julius\ Caesar\ was\ not\}.$$

The likelihood agreed upon E_1 will then be the countable measure of the historians having chosen E_1, which may be worth 0.8.

However, the reality is such that $P(E_1) = 0$ or $P(E_1) = 1$.

It can then be said that likelihood is not always a direct measure of the occurrence of a studied event, even though certain events that are associated with it are measurable. There are techniques for expert evaluation, called Delphi methods [38], to help solve these types of problems.

Probability as a measure of likelihood is associated with decision or decision-making statistics [36], whose objective is, from the structuring of data processing, in part subjective, to optimize a decision in the presence of uncertainty, and therefore risk.

2.3 Return Period of an Event

2.3.1 Notion of Quantile

Let us initially recall that [3]:
   a) for a function f to be considered as a probability density of a random variable X, it is necessary to verify, on its definition domain D, the following equality:

$\int_D f(x)\, dx = 1$

   b) for D = [a, b], the relationship between probability density function and cumulative distribution function (or non-overtaking distribution function) is:

$F(x) = \int_a^x f(x)\, dx$
$F(a) = 0 \text{ e } F(b) = 1$

By definition, a quantile x is one of the realizations of a random variable X. More precisely, a quantile of order p, noted as x_p, is the quantile associated with the probability p of X not overtaking the value x:

$$p = P(X < x_p) = F(x_p)$$

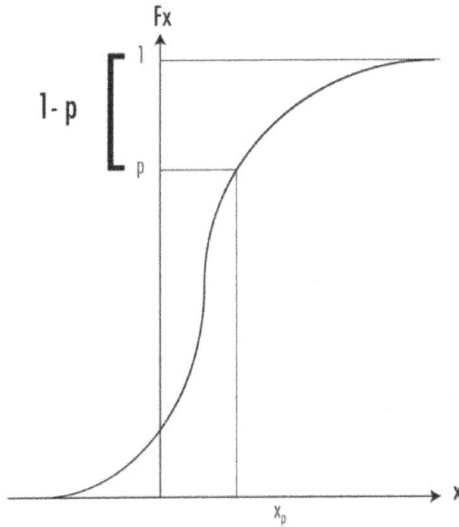

Image 2.3.1-A) Quantile x_p of a Random
Variable X

Particular attention should be paid to this concept, since some authors associate x_p with the probability of overtaking $p = P(X>x)$.

2.3.2 Return Period of a Quantile

The return period (or recurrence period) T(E) of an event E is the average time interval between two occurrences or two observations of an overtaking of x_p quantile of a random variable X. From the distribution function F(x) or the probability density f (x), the return period is formalized by [11]:

$$T(E) = 1 / f(E) = 1 / (1-F(x_p))$$

Events may be punctual (instantaneous) or need a duration of observation D (hour, ..., year) on a continuous time basis. In the latter case, the return period T will be a multiple of D.

This means that on average it is observed a realization of X greater than x_p at each T time units, which does not imply that in a given period $\Delta t = T$ there is no observation of $X> x_p$ or that several cannot occur.

The notion of return period integrates the hypothesis that there is no change in the population parameters from which observations are made that allow to estimate P(X), which is practically impossible to demonstrate in case of long periods of recurrence (or small probabilities, which is the same). This concept is closely related to the definition of probability as a countable measure: its use within contexts that rely on other definitions of probability must be careful because it can lead to conclusions with poor objective meaning.

2.4 Approximations and Errors

2.4.1 Poincaré's General Formula

Poincaré's classical formula [3] defines the algorithm for calculating the probability of the union of any events. Let $S = \cup_{i=1,n} E_i$ be the union of n events E_i on which is made no hypothesis, we have that:

$$P(S) = \Sigma_{i\text{-}1,n}\, P(E_i) - \Sigma_{\forall i,j}\, P(E_i \cap E_j) + ... + (-1)^{n-1}\, P(\cap_{i=1,n} E_i)$$

Only the knowledge of the respective probabilities of events E_i is not sufficient to calculate the probability of their union. In fact, it is necessary to calculate composite probabilities, calculated from conditioned probabilities, which express the degree of dependence among events:

$$P(E_i \cap E_j) = P(E_i) \cdot P(E_i | E_j) = P(E_j) \cdot P(E_j | E_j)$$

2.4.2 Particular Cases of Poincaré's Formula

Two particular cases of the Poincaré's formula can be identified, implying simplifications in its form:

  a) <u>Mutually excluding events</u> (or incompatible): when the occurrence of one of them implies in the non-occurrence of all the others, that is to say:

$$E_i \cap E_j = \varnothing \qquad \forall_{i,j}$$

what implies in the disappearance of the composite probabilities of the formula, that is noticeable in its reduced form:

$$P(S) = \sum_i P(E_i)$$

b) Independent events: when the occurrence of one has no influence on the occurrence of others, that is to say:

$$P(E_i | E_j) = P(E_i) \text{ e } P(E_j | E_i) = P(E_j)$$

this implies that the composite probabilities decompose into a product of the probabilities of considered events, Poincaré's formula becoming:

$$P(S) = \sum_i P(E_i) - \sum_{i,j} P(E_i).P(E_j) + ... + (-1)^{n-1} \prod_i (E_i)$$

that can also be expressed through Morgan's rules for operations with sets [3] by

$$P(S) = 1 - \prod_i [1 - P(E_i)]$$

In systems safety it becomes necessary to carefully analyze events of type (B | A), which correspond to a common cause failure or to the propagation of failures. In the first case, a common cause failure mode reduces the effectiveness of redundancies. In the second case, an aggression or particular failure implies a "cascade" failure of redundant elements or elements downstream of the functional chain analyzed.

2.4.3 Simplifications to Poincaré's Formula

It Is possible to use the reduced Poincaré's formula if the elementary possibilities are small and if the number of considered events is not very large. For example, for n identical elements of probability p, the use of the reduced formula is acceptable if np <0.1. In fact, for independent or incompatible identical elements:

$$P(S) = 1 - (1-p)^n = 1 - e^{-np}$$

because $\quad 1 - p = e^{-p} \quad$ for $p < 0.1$

likewise $\quad 1 - e^{-np} = np \quad$ for $np < 0.1$

where $\quad P(S) = np \quad$ for $np < 0.1$

This result is immediate for n identical incompatible events.

2.4.4 Accumulated Frequency

Considering an undesired event E related to the functioning of a facility whose probability of occurrence is constant in time and equals p, then the following events can be calculated:
a) Occurrence of E in the nth year:

$$P_n = (1 - p)^{n-1} \cdot p$$

b) Occurrence of E at least once over n consecutive years:

$$P_n = 1 - (1 - p)^n$$

c) Occurrence of E between year m and year n:

$$P_{n-m} = P_n - 1 \; P_n = (1 - p)^m - (1 - p)^n$$

It is important to point out that the fact that it is false to state that an event whose annual probability is "a priori" equals p will be realized with a probability equal to 1 (certainty) along $n = p^{-1}$, because of the following possible approximations for the calculation of P_n:

a) for p>0.1: there is no possible approximation of the basic formula; therefore, for $n = p^{-1}$ the latter should be used directly:

$$P_n = 1 - (1 - p)^{1/p}$$

b) for p<0.1: the first approximation of the basic formula allows to express:

$$(1 - p) = e^{-p} \Rightarrow P_n \approx 1 - e^{-np}$$

for $n = p^{-1}$ it is possible to directly calculate:

$$P_n = 1 - e^{-1} = 0.632$$

note that by the exact formula, for p= 0.1 and n=10, it is obtained 0.6513; this approximation tends to underestimate the probabilities P_n;

c) for p<0.1 and np<0.1: the second approximation of the base formula (exact) allows to express:

$$e^{-np} = 1 - np \Rightarrow P_n = np$$

In this case, it is possible to calculate P_n by $n = p^{-1}$, because this last approximation is only valid for np<0.1 and not np = 1; this approximation tends to overestimate the probabilities P_n.

2.5 Reflections About Fixation of Probability Minimum Limits

2.5.1 Preliminary Considerations

This item introduces a set of reflections on very small probabilities, associated rightly or wrongly, with events "judged" almost impossible. These probabilities appear in different stages of safety analysis:

a) in definition and demonstration of target fulfillment, where probabilities of a subjective nature are explicitly associated with undesired events and with conditions of the system environment to be considered as requirements;

b) in the choice of scenarios considered "a priori" that may lead to the undesired event, "judged" by a "decision maker" as the most plausible among the set of scenarios identified or imagined by analysts;

c) in the "post-evaluation" of probability of these scenarios, which make appear almost negligible values to determine events combinations.

These three stages of analysis and assessment correspond to three decision stages whose suitability has a direct impact on the estimated level of safety of the system. The question that arises for these three categories of probabilities can be formulated as follows:

"From what limit of probability or unlikeliness can the events or scenarios of identified events, therefore known, be ignored in security analyzes, and then ignored in decisions that could take them into account?"

2.5.2 Credibility of Safety Objectives

A safety target [39-41] is defined by two parameters:
   a) the description of the undesired event, which may be described by a limit of damage;

   b) the frequency or verisimilitude associated with this undesired event, in units consistent with its description.

The probability setting in the target is linked to the severity of consequences of the undesired event during the activity under consideration. It follows that the probability setting is based on the observed frequency of a natural or technological event which has similar consequences.

The credibility of safety of a system is directly linked to the level of safety aimed at and demonstrated, which can be defined through:
   a) the ambition of the safety target associating an unacceptable damage with a very small probability, out of possibilities of observation;

   b) the confidence in demonstrating the achievement of the target through a set of tasks well-identified and clearly described of the safety analysis.

Whenever possible, in a more or less simple way, defining and issuing safety targets, ensuring the level of security reached by probabilistic techniques, from the available data, it proves to be a much more complex task. In fact, this last activity is based on the acquired confidence during studies and actions undertaken during design, development, manufacture, construction and assembly and operation of current and previous similar systems.

This poses the real problem of credibility of demonstrated safety targets, which corresponds mainly to the effectiveness of security actions validated by their experimental verification at the considered level (elementary system, subsystem, component, item). However, this validation may be materially impossible, or unacceptable, considering the damages that would result. The material impossibility may arise not only from the considerable number of tests to be carried out to allow observation of the undesirable event defined by the target, but also from the fact that the validation is out of the possibilities of experimentation, for example, due to the considered time scale considered , which may be several centuries.

Two examples illustrate the difficulties of conducting an adapted validation experiment:

a) the regulation of air transport associates an objective probability, ranging from 10-5 and 10-7 per flight hour, with a situation which involves "a significant reduction in safety margins achieved by a difficulty of the crew in dealing with unfavorable situations that may lead to injuries to passengers"[40];

b) the regulation of large dams requires that the bleedings allow the runoff of a ten-thousand-year flood (return period of 10,000 years) without damaging the dam.

The return periods of these two events are respectively of the order of 100 years and 10,000 years, which in both cases is out of experimental possibilities, not to mention the problems related to the experimentation itself.

Consequently, the direct demonstration of the binary type ("no event will be observed in n years") not being possible, validation of compliance with the target is largely associated with the validation of efforts in methodological and technical terms described in the safety analysis.

If in certain cases, safety level assessment can be done at the system level using extreme value laws, most safety analyzes perform accident scenario modeling, whose credibility goes through:

a) the representation of the models that correspond, in particular, to the completeness defined by the number of parameters or variables and by laws governing the relations between the internal and external variables to the system; and

b) the credibility of data.

This implies a "natural" uncertainty (and, more generally, a lack of knowledge) about the demonstration procedure, its results and its interpretation. But this uncertainty should not however be compared to the uncertainty associated with assertion, without sufficient study, of "a priori" impossibility of occurrence of an undesirable event defined in the target.

Indeed, in the first case, appropriate measures have been studied and proposed, and in the second case the lack of measures stemming solely from the statement "this cannot happen" leaves a real state of insecurity that could lead to a catastrophe that could be avoided.

2.5.3 Selecting Scenarios for Analysis

Classically, the identification of scenarios leading to an undesired event passes through experience and imagination of those who develop the Preliminary Risk Analysis of the system under consideration.

Scenarios resulting from this analysis are, in general, not weighted or graduated by preliminary assessments. The only hierarchy is made according to four classes:

    a) C1: Scenarios already observed, judged realistic;

    b) C2: Scenarios already observed, but judged unrealistic, taking into account the prevention measures already taken;

    c) C3: Scenarios not observed, but considered realistic;

    d) C4: Scenarios not observed, and judged unrealistic.

The quality of the judgment between realists and non-realists is intimately linked to the depth of the body of knowledge of the set of analysts, complemented by that one of the decision maker, who has a preponderant weight, due to its function in the organization. Indeed, the decision maker's dilemma can be summed up as:

    a) accepting the consideration of a scenario whose occurrence, although possible, is "a priori" unlikely during the life of the system; this consideration may unnecessarily penalize the project by introducing additional technical, economic or operational requirements;

b) rejecting the consideration of a scenario whose occurrence is considered "a priori" as hard to imagine during the life of the system and then accepting, implicitly or explicitly, its consequences; this decision does not penalize the project, but can generate additional costs of operation, and may even lead to the premature decommissioning of the system as of the first incident judged socially unacceptable.

It should be pointed out that, according to the considered risk component, probability or severity, the decision maker will lean towards one or the other option:

a) if he only considers the small verisimilitude of the scenario, he will reject it, whatever its gravity, which is a decision based on the short term; or

b) if the severity of consequences is considered before everything, the scenario will be accepted independently of its verisimilitude, which is a decision based on the long term.

Under uncertainty, a simplistic decision rule to consider scenarios consists in associating them "a priori" with a limit of verisimilitude from the target associated with the undesired event under consideration. This limit can be fixed by working with the hypothesis that there should be no more than 100 possible scenarios, identified or not, leading to the undesired event. This implies considering only scenarios whose verisimilitude is two orders of magnitude less than the undesired event.

Thus, for a 10^{-3}/year target, the analysis shall be limited to scenarios whose probability is greater than 10^{-5}/year. On the other hand, for a 10^{-7}/year target, scenarios (possibly formed by the conjunction of several events) or rare events of probability equal to or greater than 10^{-9}/year will be considered. Obviously, such a rule cannot be applied without prior analysis, even if succinct.

2.5.4 Absolute Limit of Negligible Probability

The last example presented raises a fundamental question about the existence of a minimum limit of negligible probability, that is to say, to determine the negligible event and whose probability p serves as the lower truncation terminal in calculations and probabilistic safety assessments.

It is possible to look for a despicable envelope event, and considering natural events that are abstractly possible, but which are not taken into account in the normal life of men and societies, such as the so-called "end of the world."

Defining e as the survival of the Universe after N years and knowing that it exists for n years, it is possible to associate e with a probability q that formalizes from its duration of life or existence T whose classical expression is given by:

$$P(T{\geq}n, T{\geq}N) = P(T{\geq}N) \text{ with } n < N$$
$$= P(T{\geq}n) \times P(T{\geq}N \mid T{\geq}n)$$
$$\text{with } P(T{\geq}N \mid T{\geq}n) = q$$
$$\text{then } q = [P(T{\geq}N) / P(T{\geq}n)]$$

The following hypotheses are then made:
a) the annual probability of disappearance of the Universe is constant and equals p:

$$P(T{\geq}N) = (1-p)^N$$

b) the universe was created at $n = 1.5 \cdot 10^{10}$ years;

c) $P(T{\geq}n) = 1$, since it is observed that the Universe survives today; and

d) Arbitrarily, there is a one-to-two chance that the universe will continue to exist in the next year. It is possible then to calculate p:

$$(1-p)^N = 0.5$$
$$p = 4.6 \cdot 10^{-11}, \text{ that is, } p > 10^{-11}/\text{year}$$

To verify the sensitivity of p to the arbitrary hypothesis d), we could verify:

> d') there is one chance against 2 that the Universe survives to:

$$N = 2n = 3 \cdot 10^{-10}$$
$$p = 2.3 \cdot 10^{-11}, \text{ that is, } p > 10^{-11}/\text{year}$$

> d") there is one chance against 99 that the Universe survives to:

$$N = n + 1 \text{ years}$$
$$p = 10^{-12}/\text{year}$$

We find that the results have the same order of magnitude, and therefore are not sensitive to the arbitrary hypothesis d).

Knowing that nothing can be more catastrophic than the disappearance of the Universe, the minimum limit of negligible probability can be set between 10^{-11} and 10^{-12} per year, or 10^{-15} to 10^{-16} per hour. Consequently, it can be considered as unrealistic, if not totally absurd to take into account and manipulate, within safety analyses, probabilities equal to or inferior than this limit.

3. Formalizing the Concept of Risk

3.1 Definition and Concept

3.1.1 Origins of Risk

The origins of risk correspond to two complementary aspects: energy management and the nature of the parties involved. Given a system S, in order to assure its mission, it must necessarily:

a) possess energy in the most general sense of the term (electrical, mechanical, chemical, magnetic);
b) be able to acquire energy; and
c) be able to filter out or eliminate all surplus energy introduced by a set of internal or external aggressions of different natures.

Inadequate design, construction, integration or operation of this system inevitably implies an increase in the level of uncertainty, which may exceed an acceptable limit corresponding to a set of planned measures and means, which can then lead to an accident. This uncertainty has three generic causes:

a) possibility of occurrence of known random events, but whose occurrence moments are unforeseen, such as failures, human errors, external aggressions or threats in the general sense of the term;
b) partial or total lack of knowledge of the failure modes of parts of the system; and
c) partial or total lack of knowledge of the functioning mode of the system in particular circumstances, for example, an unforeseen or unknown natural environment.

As has been seen previously, uncertainty must not be assimilated to ignorance: it must, on the contrary, be the starting point for oriented research actions and appropriate reactions. The very nature of parts of the system S or elements that enter it or which it produces, or the environment within which it evolves, can be a source of danger and therefore lead to a risky situation in normal or abnormal functioning configurations. Therefore, the purpose of safety analyses of the system is:

a) to analyze the set of event scenarios leading to the occurrence of an accident whose initializing event (or origin) is one of the cited causes;

b) to quantify eventually the likelihood of these scenarios in probabilistic terms; and

c) to propose actions for risk reduction, materialized by propositions of "safety barriers", aiming to control or limit the evolution of a dangerous scenario;

Aiming at its completeness, risk analyses should systematically explore the two states of functioning of the system S:

a) normal (specified), to identify intrinsically dangerous elements, internal and external to the system (environment);

b) abnormal, to identify internal failures that can lead to an accident, all causes confused.

3.1.2 Nature of Risk

As it was settled, the relative risk of occurrence of an undesired event during a dangerous activity is determined by two parameters:

a) <u>probability</u> of occurrence of the undesired event (probability of causes); an

b) <u>severity</u> of the consequences which ultimately correspond to mission failure, deaths, serious injuries, destruction of property, degradation of the environment.

It can be said that the risk related to an undesired event, an event that is considered in the <u>present</u>, is defined simultaneously:

a) by a parameter describing in a synthetic way a sequence of events from the past (probability of occurrence of the set of causes);

b) by a parameter describing a set of events potentially observable in the future (severity of consequences).

Two types of risk can then be considered:

a) average risk during a given activity, defined as accumulated risk that exists during the execution of the activity under consideration, for which the unit of time is not explicit, since it corresponds to the duration of the activity;

b) instant risk, defined as permanent risk that exists during the activity under consideration.

For example, if r(t) is the instant risk to the activity under consideration, whose duration is T, it can be considered as the average risk of this activity:

$$R = \int_T r(t) \, dt = r_m \cdot T$$

It is also possible to define the same way the average risk to which a certain population is exposed during an activity, from the number of its elements that can suffer injuries. In case of a human population, there is a personal risk to be considered, which refers to a single individual.

3.2 Gravity of Consequences

3.2.1 Incidence of Consequences

The consequences associated with an undesired event affect people, the environment and assets and properties.

At least six human groups can be identified which could be increasingly affected by potential consequences of normal,

abnormal and accidental operation of an industrial facility:
a) facility workers directly affected by dangerous activities;
b) facility workers not directly affected by dangerous activities;
c) residents of communities neighboring the facility;
d) residents of the socioeconomic region in which the facility is located;
e) residents of the administrative macro-region in which the facility is located;
f) public in general that despite residing a long distance from the facility may suffer consequences through effects on the environment.

At least six geographic areas (or ecosystems) can be identified which could be increasingly affected by potential consequences of normal, abnormal and accidental operation of an industrial facility. It should be noted that these regions are closely related to the human groups mentioned above:
a) physical area of the facility where dangerous activities are carried out;
b) physical limits of the facility;
c) local ecosystem;
d) regional ecosystem;
e) macro-regional ecosystem;
f) global ecosystem.

3.2.2 Classification of Consequences by Types of Manifestation of their Effects

The effects of the consequences of an undesired event may manifest in relation to the time elapsed after the event in:
a) immediate effects;
b) delayed effects;
c) long term effects.

In addition to the time lag, the effects can be expressed in the following ways:

a) deterministic, that is to say, given the occurrence of the event, there is certainty of occurrence of the effects, usually immediate; and

b) stochastic, that is to say, given the occurrence of the event, there is a certain probability or an increase in the previously existing probability of occurrence of the effects, usually delayed or in the long term.

Typical examples of stochastic effects are the biological effects of ionizing radiation [95,96] and the biological effects of toxic and/or carcinogenic chemicals, both at low doses (below a deterministic threshold).

Among these effects, cancer may be the most representative: there is a high natural individual risk, independent of any human industrial activity, of a certain person contracting cancer throughout his life. The exposure to ionizing radiation or to certain chemicals in moderate doses increases the likelihood of this effect, which however remains stochastic.

Considering the high natural likelihood of occurrence of cancer throughout the life of a normal individual, the order of 25%, it becomes impossible to demonstrate, in a deterministic way, the cause and effect relation for a specific case of cancer. Such demonstration can only be done statistically and for a large number of cases occurring in individuals of a certain exposed population, through epidemiological methods.

It can be concluded that the risks associated with the effect of cancer occurrence are doubly probabilistic [97], that is to say, there are:

a) a probability that a human group will be exposed; and

b) a probability that, at time t after exposure, there will be n additional cases of cancer (beyond those due to natural causes) among individuals in the group.

3.2.3 Classification of Consequences by Gravity Class of Their effects

Regardless of probability, the risks can be qualitatively classified into four categories related to their gravity:

a) Catastrophic Risk, which corresponds to consequences such as irreversible damage to man (death, permanent disability) and total destruction of the system and/or its environment; In principle, considering the phenomenon range or the available time, no safety action can be foreseen or even "a priori" be really effective; consequently, the identification of a catastrophic risk must systematically imply the search and validation of prevention actions; for these actions, a catastrophic event must become rare, that is to say, of very small likelihood; in addition, it is imperative to identify the initiating event(s), called "precursor(s)", whose occurrence is almost deterministically associated with the catastrophic event;

b) Critical Risk, which corresponds to consequences such as reversible damage to man (serious but not permanent injuries), to the system (partial destruction) and to the environment; an emergency (or safeguard) procedure should prevent such consequences from occurring; this implies that the identification of a critical risk should lead to the search and validation of prevention and protection actions;

c) Significant Risk, which corresponds to consequences such as minor injuries, failure of the mission without destructing the system, or long unavailability;

d) Minor Risk, which corresponds to failure of system elements with no consequences on the success of the mission nor on safety.

The first two classes of risk are related to safety. The last two are usually related to the success of the mission.

3.3 Determination of Safety Objectives

3.3.1 Acceptable Risk

Acceptable (or permissible or limiting) risk is the risk resulting from an explicit decision established in an objective manner by comparison to known and currently accepted risks, from natural, social, technological or economic causes [42, 43].

However, the acceptability of risk by individuals and by society is influenced by a number of factors [44]. The most important of them was evidenced by Starr [43], and it is linked to the voluntary or involuntary character of the incurred risk: it is accepted to incur voluntary risks up to three orders of magnitude higher than involuntary risks. There are even other factors [45], such as immediate or delayed consequences, the presence or absence of alternatives, the precise or imprecise knowledge of the risk, the common or particular danger to certain people, the reversibility or irreversibility of consequences.

In industrialized countries, the risk of death due to diseases is in the order of 10^{-2}/year, which is a benchmark for the highest levels of risk that are unintentionally accepted. The lowest level of unintentionally accepted risks is the one that results from exceptionally severe natural phenomena, which is in the order of 10^{-6}/year. Between these two edges, the public seems to accept involuntary risks in terms of the benefits obtained [44,45].

An American research [46] schematically came to the following

conclusions about annual levels of individual risk of death:
a) 10^{-3}/year: this level of risk is unacceptable, that is to say, as soon as the risk approaches this level, immediate actions must be taken to reduce it.
b) 10^{-4}/year: the public claims public expenditure to control and reduce these risks (highways, fires);
c) 10^{-5}/year: the risks at this level are identified by the public, that however is satisfied with certain general rules to reduce them; and
d) 10^{-6}/year: the risks at this level do not disturb the average individual; he recognizes these accidents but thinks that "this can only happen to others", showing resignation in the light of this risks.

The authors of this research conclude that the individual risk of death of 10^{-7}/year is an acceptable upper limit for the risk of nuclear power plant accidents.

There are also numerous factors that affect the perception of risk by the public. Another interesting British research [47] has shown the relationship between risk perceived by the public and real risk. The authors came to the following conclusions:
a) causes of death of "illness" and "road accidents" type are considered as equivalent, when in fact the first ones are ten times more numerous;
b) the risks that significantly contribute to the total number of deaths are underestimated; and
c) the risks that little contribute to the number of deaths are overestimated; so, exceptional events are considered to be much more deadly than they really are.

This poll confirms the "common sense" that the public judge "less dangerous" an activity that makes 1 dead every day than another that makes 365 deaths in a single day of the year. Risk perception depends on numerous moral and psychological factors that are difficult to quantify or even to explain.

When referring to human death statistics, it should also be borne in mind that they vary significantly from time to time, between geographical regions and by age group of the population concerned.

Definitely, it can be concluded that acceptable risk, which is the projection of the collective, social or economic perception of the associated danger, cannot be defined in a universal way [48].

The search for the acceptable risk value is a compromise between what the responsible body is willing to pay, if it takes into account "a priori" the occurrence of risk and resulting safety measures, and how much it would have to pay "a posteriori", if hypothetically the risk is ignored, considering:
   a) the costs of reparations for human, material and environmental damages;
   b) the unavailability costs;
   c) the impact on the media,

factors that can lead to the definitive stop of the activity.

From this perspective, the overall cost of security must be considered, which includes:
   a) cost of studies and safety devices, which are "a priori" costs;
   b) cost of accidents, which are "a posteriori" costs;

An economic optimum can be determined based on:
   a) investment costs associated with residual levels of insecurity (or risks) evaluated "a priori";
   b) costs of residual insecurity (or risks) associated with the costs of repair that can be induced "a posteriori".

Representing in abscissa the residual insecurity levels corresponding to the accident probability, the determination of the economic optimum can be visualized by the Image 3.3.1-A.

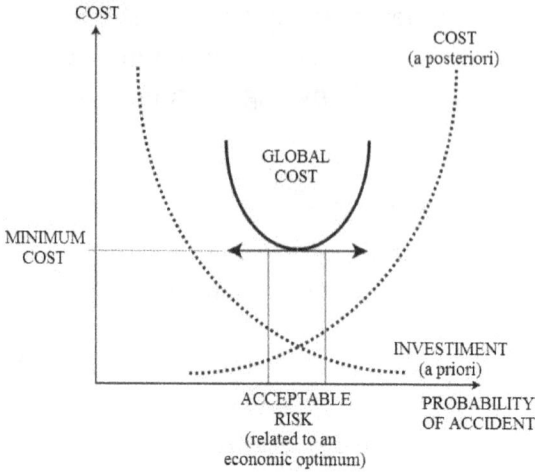

COST

COST
(a posteriori)

GLOBAL
COST

MINIMUM
COST

INVESTIMENT
(a priori)

ACCEPTABLE
RISK
(related to an
economic optimum)

PROBABILITY
OF ACCIDENT

Image 3.3.1-A) Safety Costs Optimization

3.3.2 Definition of General Safety Objectives of the System

Choosing the General Safety Objectives (GSO) is a fundamental step in Operational Safety of systems because it bridges the boundary between the acceptable risk domain and the unacceptable risk domain of the system functioning states:

a) the setting of safety targets results from a decision of political and economic authorities of a higher level, and not from the safety specialist in charge of an activity or project;

b) the system target must be beyond the reach of observation or, more precisely, of direct experimentation so that it can be considered as valid; this aspect gives it a subjective nature that causes a near impossibility of demonstrating its initial care and maintenance throughout the life of the system.

The independence between the political ("decision maker") and the technical or economic one is fundamental. Indeed, for many reasons, the last two can minimize the real risks by overestimating the technological level of the system, leading to an "almost religious" (therefore ignoring its lack of knowledge domain) belief, or simply for "obscure" and less ethical reasons.

The expression of a safety target [40, 41, 49,94] associated with an undesired event contain three elements:
  a) definition of the base environment to be considered;
  b) precise definition of the undesired event;
  c) definition of an acceptable likelihood or verisimilitude of occurrence of this event.

The base environment (mission, natural and technological) contain:

  a) definition of the mission profile, that is to say, the use conditions of the system (duration, configuration) and the characteristics of the conditions under which this use is made;

  b) inventory of potentially dangerous conditions;

  c) active base environment, corresponding to a set of aggressions or non-nominal loadings in project defined by the level of loading n and the probability that this load is exceeded $P_n = P(N> n)$.

The precise definition of the undesired is made from:
  a) definition of the system and its intrinsically and potentially dangerous elements;
  b) inventory of undesired events.

The description of the undesired event must be clearly expressed, whether by the total or partial duration of the dangerous activity, or by the time unit of exposure to the sources of danger, or by the specific use of dangerous equipment.

Due to the random nature of safety targets, the acceptable probability of occurrence of the undesired event can be qualitatively or quantitatively defined.

3.3.3 Qualitative Safety Objectives

Qualitative targets can be expressed by a likelihood scale [50] or from a project criterion connected to the operational phase:

a) Likelihood Scale: established from arbitrary levels such as:

- Impossible

- highly unlikely or highly rare

- unlikely or rare

- plausible

- very plausible

- almost surely

- surely

Note that the qualitative of impossible and surely refer to events belonging to the certainty domain, while the others belong to the uncertainty domain; the relation between this scale and the previously defined scale of gravity can be established by imposing that a catastrophic event is highly unlikely and that the occurrence of a critical event must be at least unlikely; it is evident that such a subjective scale varies from one activity to another, even though the events associated with extreme likelihoods are in principle common;

b) Project Criteria [51] such as:

- "Fail Safe", for which any failure leads to a safe state; the system is then said to be "intrinsically safe" or has "integrated safety";

- "Fail Operational / Fail Safe", for which the first failure allows to continue the mission in safety and the second failure leads to a safe breakdown state;

- "No fail", for which no failure regarding the function or considered equipment is acceptable.

Another project criterion is the principle of defense in depth or principle of three barriers, used especially by the nuclear industry, but which can be generalized to other activities.

3.3.4 Quantitative Safety Objectives

The definition of quantitative targets is classically made from statistical observations on the occurrence of events of the same gravity considered as admissible, as seen previously. It is the case of the natural death probability that is generally considered as a base for catastrophic events.

For an undesired event of consequences with multiple gravities, that is to say, affecting different populations, properties and environments, it is necessary to associate to each of the consequences a probability associated with each specific risk. This is the case of the material and technical undesired events presented by the classic report on nuclear power risks [41], in which many values of number of deaths were associated with the odds of corresponding scenarios.

3.4 Representation of Risk and Safety Objectives

3.4.1 Description

The risk, defined from two components, can be visualized on a graphic whose abscissa represents the scale of gravities, and the ordinate represents the scale of probabilities (frequency or admissible likelihood).

This criticality graphic, called Farmer's Diagram [26], allows a view of acceptable and unacceptable risk domains. Chart 3.4.1-A is an example of Farmer's Diagram.

Chart 3.4.1-B presents the establishment of criteria on a Farmer's diagram. The boundary between domains is defined by the curve joining the pairs (g1, p1), (g2, p2) and (g3, p3). The current trend is, however, to consider three domains, including an intermediate domain within which actions to improve safety should be analyzed under an optimization approach (ALARA procedure), taking into account economic and social factors.

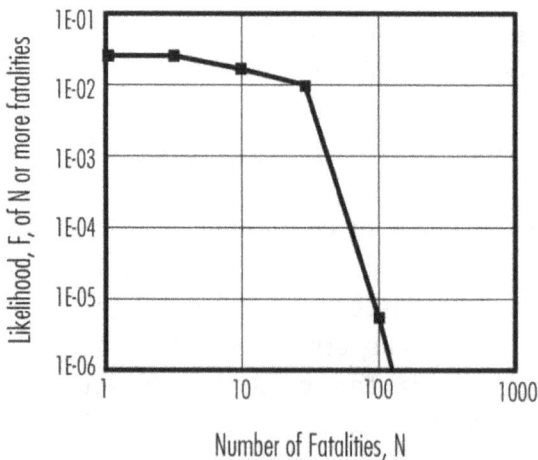

Chart 3.4.1-A) Typical Farmer's Diagram

Another classic representation of risk and safety targets is in the form of a table called the criticality table [39], of which Table 1.1.4.1-C is an example.

Farmer's diagram has the advantage of visualizing a gravity distribution function, being therefore closer to the definition of risk and also wealthier in information.

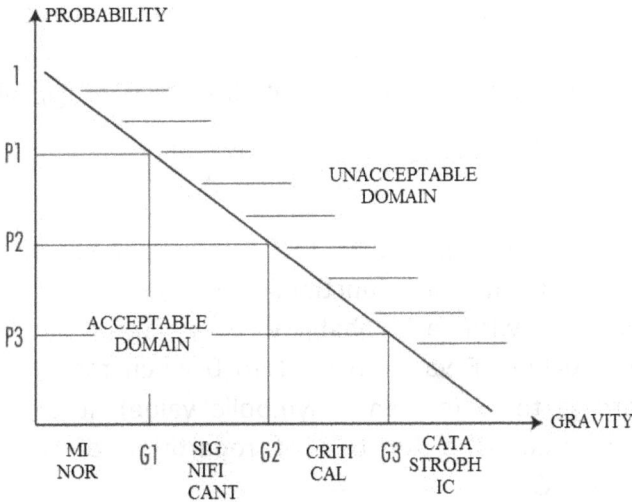

Chart 3.4.1-B) ALARA Farmer's Diagram

GRAVITY PROBABILITY	Minor	Significant	Critical	Catastrophic
almost certain	acceptable	unacceptable	unacceptable	unacceptable
likely	acceptable	acceptable	unacceptable	unacceptable
unlikely	acceptable	acceptable	acceptable	unacceptable
rare	acceptable	acceptable	acceptable	acceptable

Table 3.4.1-C) Typical Criticality Table

3.4.2 Nature of Representative Risk Curve

As it was seen, it is accepted for a given activity (or system):
a) occurrence of an event of insignificant gravity with a probability close to 1; and
b) occurrence of a severe gravity event with a non-zero probability, but much lower than 1.
c) This way, if we consider the function $p=F(g)$, which is non-zero in the gap of the considered gravities G, there is a lower terminal g_0 of gravities such that $F(g_0)=1$ can be stated. Next, it is possible to calculate the integral of F over $[g_0, \infty[$ and it is shown that:

$$\text{d) } \int_G F(g)\, dg > 1$$

e) It is concluded, therefore, that F cannot be considered as a probability density function. However, it is logical to associate with a probability $p = F(g) \ll 1$. Since by construction, F varies from 1 to 0 when the gravity goes from g_0 to ∞ (evidently symbolic value), it can then be considered as a function of repartition of the random variable G.
f) Let f be the probability density of the variable G, by definition the derivative of the function repartition of G:

$$\text{g) } F(g) = P(G>g) = \int_G f(g)\, dg = p$$

h) In particular, it is found that:

$$\text{i) } F(g_0) = P(G> g_0) = \int_G f(g)\, dg = 1$$

j) Consequently it can be concluded that the criticality curve has the same nature of a probability repartition function and that it is therefore not accurate to state that a gravity risk g should have a probability equals p, but that a gravity risk g must have <u>at most</u> a probability equals p.

3.4.3 Average Gravity and Objective Average Risk

From presented notations, it is possible to calculate:
a) <u>Average Gravity</u>:

$\mu_G = \int g. \ f \ (g) \ dg =$
$= \Sigma_i \ g_i . \ P(G=g_i) =$
$= \Sigma_i \ g_i . \ f(g_i)$

b) <u>Objective Average Risk</u>:

$\mu_R = \int g. \ F \ (g) \ dg =$
$= \Sigma_i \ g_i . \ P(G \geq g_i) =$
$= \Sigma_i \ g_i . \ F(g_i)$

To formalize the concept of risk, it is then possible to indifferently use either average gravity or objective average risk, given that there is a direct relation between them. The criticality curve, however, allows us to directly calculate the objective average risk. For the sake of simplicity, the objective average risk is calculated for a given risk class R_i of gravity G_i:

$$\mu_{Ri} = G_i . \ P(G \geq G_i) = a = cte.$$

the iso-risk curve is taken as the criticality curve. It is deduced then that: $\qquad \mu_R = \Sigma_i \ \mu_{Ri} = n.a$

As Farmer's diagram show - Chart 3.4.3-A, for which there is an objective average risk equal to 10^{-3} for each of the three considered risk classes, which implies a total objective average risk of 3.10^{-3} $.

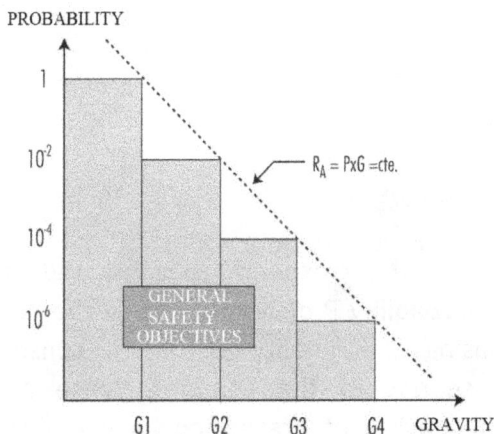

PROBABILITY

$R_A = PxG = cte.$

GENERAL
SAFETY
OBJECTIVES

G1   G2   G3   G4   GRAVITY

Chart 3.4.3-A) Quantitative Example of Farmer's Diagram

From the gravity definition interval $[g_0, g_1, ..., g_n]$, it was defined the average gravity μ_G by calculating the barycenter of gravity weighted by its probability density (or likelihood) $[p_0, p_1, ..., p_n]$:

$$\mu_G = \Sigma_i \, p_i \cdot g_i$$

While calculating the objective average risk, the weighting is done by the probability of repartition:

$$\mu_R = \Sigma_i \, g_i \cdot \Sigma_j \, g_j = \mu_G + \Sigma_i \, g_i \cdot \Sigma_{j=i+1,n} \, p_j$$

It is concluded, therefore, that the objective average risk can give a greater weight to more severe gravities than to the average gravity. Indeed, between two distributions of the same average gravity (g_i, p_i) and (g_i, p_i), the higher the considered probabilities the bigger the difference between the respective objective average risks. This effect is even more noticeable the higher the average gravity.

3.5 Transition from Unacceptable Risk to Acceptable Risk

3.5.1 Safety Actions

Safety analyses and their resulting actions aim to eliminate, reduce or control a risk identified and recognized as unacceptable after its evaluation and comparison with its target. The actions that allow to go from of an acceptable risk are three:

a) Prevention;
b) Protection; and
c) Reinsurance.

3.5.2 Preventive Actions

A preventive measure corresponds to a risk reduction action by decreasing the probability P of occurrence of the undesired event, without decreasing the gravity G of its consequences. This definition corresponds to the common sense associated with prevention that consists of preventing an event from occurring, without systematically considering its consequences.

From an unacceptable risk identified as A in a Farmer's diagram, the preventive action consists of passing to an acceptable risk B by parallel displacement to the axis of probabilities. The theoretical safety margin is then defined between the resulting probability after the preventive measure and the objective in probability that belongs, by definition, to the criticality curve.

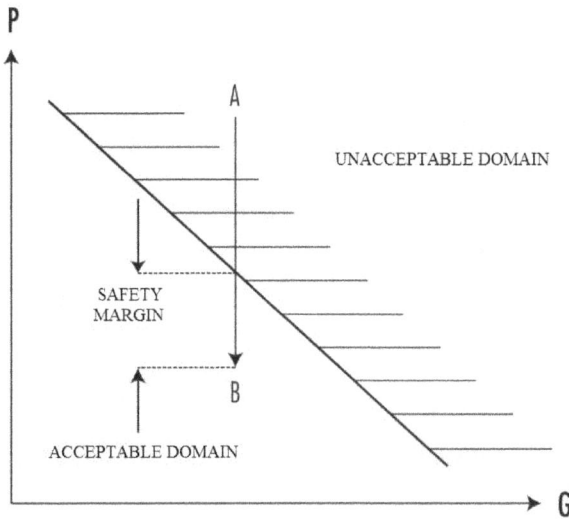

Chart 3.5.2-A) Preventive Action

3.5.3 Protective Actions

A protective measure is a risk reduction action based on reducing the gravity of the consequences G of the undesired after its occurrence, without decreasing its likelihood. This definition corresponds to the common sense associated with protection that consists in limiting the consequences of an event, without considering "a priori" its probability of occurrence, eventually close to 1.

From an unacceptable risk identified as A in a Farmer's diagram, the protective action consists of passing to an acceptable risk B by parallel displacement to the axis of gravity. The theoretical safety margin is then defined between the resulting gravity after the protective measure and the objective in gravity that belongs, by definition, to the criticality curve.

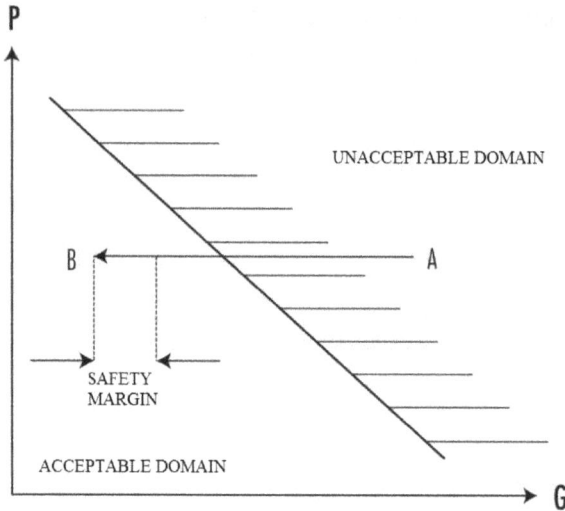

Chart 3.5.3-A) Protective Action

3.5.4 Reinsurance Actions

Reinsurance, in contrast to prevention and protection, is not intended to reduce the probability nor gravity of an undesired event. Its purpose is to transfer to a third party (the insurer) in whole or in part the financial consequences of risk.

From an unacceptable risk identified as A in a Farmer's diagram, the reinsurance action consists in shifting the criticality curve until point A is in the acceptable domain. The artificial displacement of the criticality curve has as counterpart the payment of a premium to the insurer, whose value is a function of:

a) number of "clients" that are insured against this risk;
b) probability of occurrence of undesired event, estimated from statistics;

c) protective and preventive efforts on the clients to reduce the gravity and/or likelihood of the insured risk

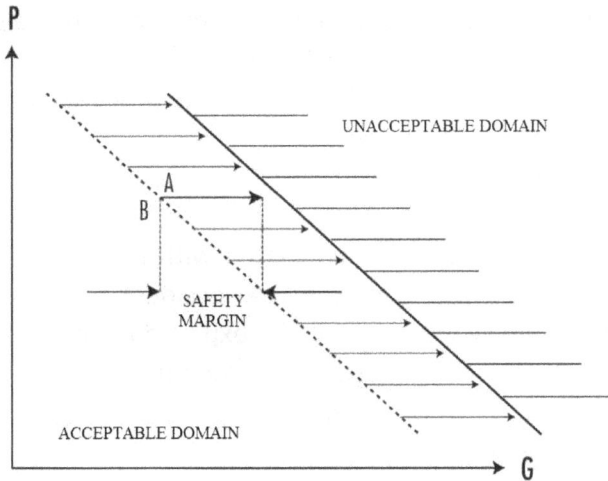

Chart 3.5.4-A) Reinsurance Action

3.6 Formalizing the Notion of Risk

3.6.1 Return Period Associated with Risk

A risk can be associated with the occurrence of an event characterized by exceeding a limit value x_p of an internal or external parameter to the given system. From the repartition function F of the random variable X, we try to estimate the frequency of occurrence p of not exceeding a given value x_p of the random variable and it is deduced its return period $T(x_p) = 1 / (1-p) = 1 / (1-F(x_p))$.

From this expression, the probability of observing, in n units of time, an event of return period of T units of time is expressed by:

$$P(X_n > x_p) = 1 - (P(X < x_p))^n$$
$$= 1 - (F(x_p))^n$$
$$= 1 - (1 - (1/T(x_p)))^n$$

For x_p and T large enough:

$$P(X_n > x_p) = 1 - e^{(n/T)} = n/T$$

Inversely, for high values of the variable, it is determined the quantile x_p corresponding to a probability of occurrence p solving:

$$P(X_n > x_p) = 1 - (G(x_p))^{1/n}$$

where $G(x_p)$ is the repartition function of the random variable X_n.

3.6.2 Empirical Average Risk

Consider an accident directly associated with the occurrence of an undesired event A. The gravity of the accident is defined by the number of victims M among a set of n exposed people. The average number of victims is then expressed by the ratio:

$E(M) = \sum_{m=0,n} P(A, M=m) \cdot m$

with: $1 \geq P(A, M=m) \geq 0$

$\sum_{m=0,n} P(A, M=m) = 1$

being: $P(A, M=m) = P(A) \cdot P(M=m \mid A)$

following: $E(M) = P(A) . \sum_{m=0,n} P(M=m \mid A) \cdot m$

being: $E(M \mid A) = \sum_{m=0,n} P(M=m \mid A) \cdot m$

getting: $E(M) = P(A) . E(M \mid A)$

This fundamental equality expresses the relation between the empirical average risk E(M), the probability of the accident P(A) and the average gravity E(M | A). It can be expressed in a simplified way by:

$$R = F. G$$

where

R: empirical average risk defined in terms of cost of consequences per unit of functioning (time or operation) which should be a requirement of the system S specifications;

F: frequency or likelihood of the accident or number of accidents per unit;

G: average gravity expressed as the average cost of consequences by accident (number of deaths, amount of losses).

Note that there is a practical difficulty to know P (M = m $|$ A). One way to get around this difficulty is to parameterize the risk by the distance of people to the source of danger.

The objective average risk presented above for a global risk is defined by a set of gravities g and associated consequence probabilities P (G> g) of a set of undesired events. The empirical average risk is defined in practice from the probability of the causes and the mathematical expectance of the consequences (average gravity).

3.7 Interest and Inconveniences of Risk Quantification

3.7.1 Interest of Probabilistic Language

Quantitative analyses are developed when someone wishes to implement safety more efficiently and to assess the level of safety of a system for a given activity. In both cases, the main tool is the calculation of probabilities, which presents numerous specific advantages:

a) it lends to mathematical treatment (addition and product, in particular);
b) it allows a better and more realistic repartition of responsibilities among teams, limiting possible interpretations;
c) it highlights the respective weights that must be attributed to the preventive and protective measures in demonstrations to be made;
d) it allows establishing a hierarchy among accident scenarios, eliminating those of low likelihood;
e) it leads, for each elementary system, to an optimization of the design effort and to a better evaluation of the level of reached and guaranteed safety, requiring a detailed analysis of safety devices and their behavior within each environment considered;

f) at last, based on the acquired results, it allows to assess better the importance of identified weaknesses of the system from a safety point of view and to accept them knowingly.

3.7.2 Limitation of the Use of Probabilistic Language

The use of probabilistic language without particular precautions can lead, however, to two major drawbacks:

a) limitation or even a decrease in the demonstrated level of safety if is considered the "absolute value" of accepted probabilities from the experience of previous systems (partial limitation of the knowledge domain);

b) an unreasonable increase in experimental expenditure, if it is desired to carry out a statistical demonstration, which by the way is extremely difficult to carry out in most cases, of the achieved level of safety.

3.7.3 Principles of the Use of Probabilistic Language

To avoid false interpretations, it is proposed that the use of probabilistic language within the methodology of safety analysis should be based on four principles.

a) during the project, and in the light of the imposed safety target (expressed as a probability p of occurrence of an undesired event per unit) a number of independent safety barriers, understood as a material, logical or human artifice, should be predicted, placed within the evolution of an event scenario to limit or interrupt its progression, and that according to its nature or position may have preventive or protective effects, such as:

$$1 > p \geq 10^{-2} : \text{at least 1}$$
$$10^{-2} > p \geq 10^{-4} : \text{at least 2}$$
$$10^{-4} > p \geq 10^{-6} : \text{at least 3}$$

b) It follows that the stated credibility and technical effectiveness demonstrated within this example for a barrier is at least 10^{-2} by undesired event or aggression; consequently, we work with the hypothesis that 10^{-4} is not credible for a barrier, since it cannot be demonstrated; clearly, the effectiveness and independence of barriers must be duly established.

c) The estimation of the effectiveness of a safety barrier should always be based on an in-depth analysis of failure modes of the system elements according to a coherent method;

d) The quantitative results provided express the experts conviction, which is justified by the estimation of the effectiveness of the safety barriers from the in-depth analysis carried out during the project and validated by suitable tests and experimental culture (studies and tests carried out during the development of the considered system and previous equivalent systems or even different ones) and application of adapted techniques (theoretical assessment, modeling, reliability assessment, extrapolation of safety margins, simulation);

e) At the system level, safety synthesis is both easier and more exhaustive as the same analysis methods are used for each of the elementary systems and their interfaces. In this case, indeed, the synthesis itself does not generate errors and forgetfulness.

3.7.4 Observations on the Use of Probabilistic Language

As has already been seen, quantification is only possible within the knowledge domain. However, in many old and new activities, the engineering associated with them may have more or less important ramifications within the lack of knowledge domain: this then prioritizes the problem of the validity of quantification.

This real difficulty, however, should not exclude "a priori" in principle all the use of probabilistic language within a considered technical activity even before the search and determination of its knowledge domain. Indeed, even if certain domains are difficult to define, a sound scientific approach to safety, at a reasonable cost, must be sought. This approach must be qualitative as a matter of priority, but it should not exclude, without scientific and technical justification, its quantitative complement. The latter can serve as an indicator, under certain validated hypotheses, for hierarchizing risks related to architectures drawn from the same logic, which would not be allowed by a strictly qualitative approach.

Finally, it is important to note that the search and use of validated methods, whatever they are, that can improve the demonstrated level of safety to maintain it, should be one of the primary concerns of the system safety officer.

4. Safety Allocations

4.1 Definition

The safety allocations [39-40, 52-54] aim to distribute the effort corresponding to the probability component of the safety target assigned to the mission (or activity) over the set of parts of the system that perform it through functional, material and operational modeling.

4.2 Basic Principles

The purpose of safety allocation procedure is to give the different teams participating in the development of a system a set of design, construction and operation requirements defined at their level of responsibility.

The Farmer's diagram describes, for each gravity class G_k, the maximum allowable probability P_k associated with it. This pair (G_k, P_k) defines one of the safety targets of the mission. For a target mission of class k , the allocation procedure consists of distributing over the architectures of the system parts, the weight in probability of its possible participation in the occurrences P_k of the considered undesired event through the identified accident scenarios that lead to it.

This procedure is done in steps, as outlined in Image 1.4.1-A, from the mission safety target, having both objective and subjective characteristics. It is objective as the diverse analyzes already carried out allowed to identify reliable elements and deterministic scenarios from a possible feedback. It is subjective because the set of individual architectures of the parts of the system are only known and proposed after the realization of preliminary allocations, which implies integrating a part of subjectivism into decisions.

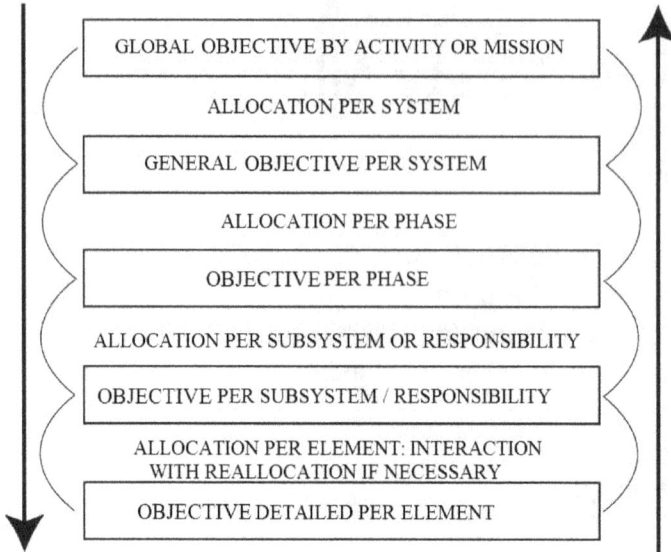

Image 4.1-A) Safety Allocation Procedure

For a given element, the main criteria that intervene in the allocation procedure are:
weight (importance) in the realization of a critical or catastrophic scenario;

a) technological complexity;
b) reliability (whether it is known or allocated);
c) characteristic: mission functional element, safety barrier, safety functional element;
d) functioning duration for the mission under consideration.

The characteristics must be established for nominal (excluding failures and external aggressions) and non-nominal (failures, external aggressions) functioning scenarios. The allocation procedure is interactive and should be performed at all levels considered in the development of the system. The type of allocation procedure can vary from level to level depending on the nature of the step and the level of knowledge.

4.3 Main methods

4.3.1 Equidistribution of risks

It consists of a preliminary allocation mode, which does not take into account the specific risks related to external interfaces and base environments.

Considering a mission formed by n phases, or a system formed by n elementary systems, or a functional or material chain formed by n elements, for a safety target p_0 at mission, system or chain level, and in the absence of any complementary information, the target allocated for each phase, elementary system or element will be p_0/n.

For example, if the duration of the system's mission is 100 hours and that the safety target associated with the occurrence of the undesired event E is $p_0 < 10^{-5}$, we will have:

a) the probability of occurrence of E must be less than or equal to 10^{-7} per hour of mission:

$$p_H < 10^{-5}/100) /h$$

b) if the analysis of the mission allows to decompose it into 10 phases, the probability of occurrence of E at any stage of the mission must be less than or equal to 10^{-6}:

$$p_\varphi = (10^{-5}/10) = 10^{-6}/ phase$$

this target would be the same regardless of the duration and criticality of the phase; in order to overcome this drawback, the average target per phase can be weighted by integrating the duration of a phase. If t_i is the duration in hours of the phase φ_i, then:

$$p\varphi l = 10^{-7}. ti$$

It should be noted that this adjustment would continue to ignore the criticality of each phase.

    c) If the system must perform 20 functions to fulfill the mission, the probability of occurrence of E after the failure of any function must be less than or equal to 5. 10^{-7} during the duration of the mission:

$$p f < (10^{-5} / 20) = 5.10^{-7} / \text{function}$$

this objective would not take into account the criticality of the function;

    d) if the system can be decomposed into 10 elementary systems, the probability of occurrence of E after a failure of a subsystem should be less than or equal to 10^{-6} during the duration of the mission:

$$p_s < (10^{-5} / 10) = 10^{-6} / \text{system}$$

this objective would not take into account the criticality of elementary systems nor their weight in the occurrence of E.

The method of equipartition of risks does not take into account the weight of each function, elementary system and phase in the occurrence of the undesired event. To overcome this drawback, three methods are considered:

    a) Weighting 'a priori';

    b) Weighting by Number of Structural Relations;

    c) Weighting by Targets or Reliability Assessments.

4.3.2 Weighting Risks 'a priori'

The weighting allocation "a priori" uses the Delphi techniques [38]. At each phase, function or elementary system an acceptable "a priori" fraction of the higher-level safety target is designated, by means of weighting coefficients V_i.

A System S carrying out a mission with a safety target equals p. If the succinct functional analysis makes it possible to identify n functions (or phases or elementary systems) to perform the mission, several types of weighting coefficients may be associated with it:

a) a_i: complexity coefficient, which globally characterizes the feasibility of the elements performing the considered function f_i; this coefficient is also associated with aspects related to innovation and different functioning conditions; the proposed scale for complexity levels a_i would be 1 (weak), 2 (medium) and 3 (strong);

b) b_i: criticality coefficient which characterizes the 'a priori' weight of the function f_i in the occurrence of the undesired event; the proposed scale for criticality levels a_i would be 1 (weak), 2 (medium) and 3 (strong).

In addition, the following hypotheses are made:

a) the more a function is potentially critical, the more it potentially intervenes in the occurrence of the accident scenario, which leads to be more restrictive in the target assigned to it;

b) one more function is complex to perform, but it is necessary to give it weight through a restrictive target.

This last hypothesis is not independent of the fact that the more a function is complex to execute, the greater its uncertainty in terms of failures and therefore the more it generates risk. This is as much more true as to the maturity of its technology.

It follows that, from the target p_0, the allocated target to the function f_i would be:

$$p_i = V_i \cdot p_0 \text{, with } \sum V_i = 1$$

$$\text{where } V_i = [(a_i / c_i) / \Sigma (a_i / c_i)]$$

4.3.3 Weighting Risks by Number of Structural Relations

The Weighting by Number of Structural Relation is based on an 'a priori' assessment of these relations, understood as a possible connection of direct or indirect interface among the elements of the system. Such a connection is the potential point of direct forwarding or after transfer of a set of internal aggressions to the system (set of critical paths).

From this assessment by phase, function or elementary system, it is possible to deduce for each one the target to be allocated. Let N_k be the number of structural relations of the phase φ_i. Making the hypothesis (that can be modulated) of equi-probability p_c of occurrence of each of them:

$$p_k = N_k \cdot p_c$$
$$p_0 = \Sigma_{k=1,n} \, p_k = \Sigma_{k=1,n} \, N_k \cdot p_c = p_c \cdot \Sigma_{k=1,n} \, N_k$$
$$p_c = p_0 / \Sigma_{k=1,n} \, N_k$$
$$p_k = (N_k / \Sigma_{k=1,n} \, N_k) \cdot p_0$$

The parameter $(N_k / \Sigma_{k=1,n} \, N_k)$ is called weighted number of structural relations.

In the absence of "a priori" information, this preliminary allocation mode is the most realistic because at each phase it already associates potential risks, regardless of its duration.

The Image 1.4.3.3-A provides an example of determining the number of structural relations of a system consisting of a manned vehicle V, a base B, a dangerous elementary system D, and a reference environment R. In this case, it is possible to identify 7 structural relations that converge on the crew.

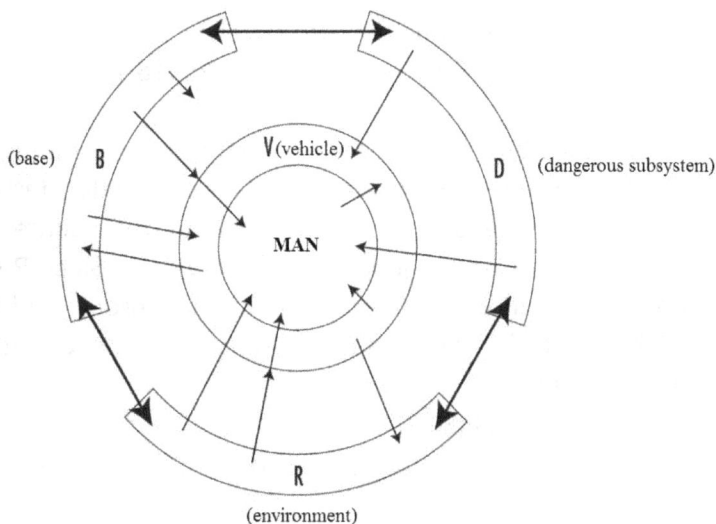

(base) B   V(vehicle)   D (dangerous subsystem)

MAN

R

(environment)

Image 4.3.3-A) Structural Relations

4.3.4 Weighting Risks by Objectives or Reliability Assessments

The Weighting by Targets or Reliability Assessments allows us to benefit safety with the efforts undertaken or to be undertaken for the technical success of the mission. This method requires the following prior information:

a) knowledge of the system and its elementary systems;

b) duration of the mission and potentially dangerous situations;

c) general safety target p_0 associated with a given accident A;

d) targets or reliability assessments R_i of n elementary systems S_i.

Let the events D (potentially dangerous situation in the considered phase), S_i (failure of the elementary system S_i among the existing n) and A (accident):

$$P(A) = P(D).\textstyle\sum_i P(A \cap (S_i \cup \bar{S_i}))$$
$$= P(D) . \textstyle\sum_I [P(S_i). P(A \mid S_i)+ P(\bar{S_i}). P(A \mid \bar{S_i})$$

There are then two types of accident configuration: resulting from an unsuitable reference environment, an under-designing performance or inadequate operation, without the failure of S_i (Type 1) and resulting from the failure of S_i (Type 2). Since P (D) is the fraction of time in which the situation is dangerous, and P (S_i) the unreliability of the elementary system S_i in the considered phase, we will have:

$$p_1 = \textstyle\sum_i P(S_i). P(A \mid S_i) = a. p_0 / P(D)$$
$$p_2 = \textstyle\sum_i P(\bar{S_i}). P(A \mid \bar{S_i}) = (1\text{-}a). p_0 / P(D)$$

where a is the likelihood ratio between type 1 and type 2 events, p_1 and p_2 are the probability weights associated with or allocated to part of the accident due to these events.

The assessment of the probabilities associated to the occurrence of type 1 events is done directly by statistical treatment. For type 2 events, a weight b_i is defined "a priori" for each elementary system, such that:

$$P(\bar{S_i}). P(A \mid \bar{S_i}) = b_i . p_2 \text{ with } \textstyle\sum_I = 1$$

where P $(\bar{S_i})$ = 1 - R_i = F_i is the unreliability allocated or evaluated from the elementary system S_i.

The safety target, conditioned by reliability, that must be allocated to the system S_i will then be:

$$q_i = P(A \mid \bar{S_i}) = (b_i / F_i) . p_2$$

b_i is the likelihood part of insecurity associated with the elementary system S_i, corresponding to the conditional probability "a posteriori" P $(S_i \mid A)$ of Bayes' formula [36]. The value of b_i is inversely proportional to the relative weight over the accident that the specialist attributes to the elementary system S_i.

The results obtained are the values of the detailed safety targets allocated to elementary systems and, by iteration, to subsystems, components and items.

To exemplify the procedure, let be a system composed of three preponderant elementary systems {S_1, S_2, S_3} with respect to the considered accident A. Being $P_0 = 10^{-5}$ per mission for the system $F_1 = 10^{-2}$, $F_2 = 5. 10^{-2}$, $F_3 = 5. 10^{-3}$ and the probability of finding itself in a dangerous situation during the mission is $P(D) = 10^{-2}$:

a) <u>Hypothesis 1</u>: the design of the system and the operating environment are well known; after that a preliminary weight of 10% (a = 0.1) connected to the use can be assigned to the occurrence of the accident under normal functioning conditions; the preliminary safety targets in normal condition will be:

$$p_1 = 10^{-4}$$
$$p_2 = 9. 10^{-4}$$

b) <u>Hypothesis 2</u>: a specialist has concluded, for example, that the accident A would have for origin three elementary systems with likelihoods $b_1 = 0.05$, $b_2 = 0.35$, $b_3 = 0.60$, implying the detailed targets presented in Table 1.4.3.4-A.

i	F_i	b_i	$b_i p_0$	q_i
1	10^{-2}	0.05	$4.5.10^{-5}$	$4.5.10^{-3}$
2	5.10^{-2}	0.35	$3.15.10^{-4}$	$6.3.10^{-3}$
3	2.10^{-2}	0.60	$5.40.10^{-4}$	$2.7.10^{-2}$

Table 4.3.4-A) Example of Allocation by Reliability Weight

5. Laws of Probability

5.1 Laws of Discrete and Continuous Variables

Probabilities are used in Operational Safety of Systems to assess the risks directly associated with an identified danger and the residual risks following the implementation of a preventive or protective action. Even if these probabilities are usually punctual, their assessment requires the knowledge, in whole or in part, of the distribution of probabilities from which they come.

According to the nature of the random variable, discrete or continuous, that intervenes in the modeling of the dangerous phenomenon, or in the risk reduction action, discrete or continuous probability distribution laws will be considered.

Among various distribution laws, we will consider those listed below as having the greatest application. The mathematical development and characterization of these laws are presented by numerous references [3, 11, 31, 55]:
   a) Laws of Discrete Variable:
   * Hyper geometric
   * Binominal
   * Pascal or Binominal Negative
   * Geometric
   * Poisson
   b) Laws of Continuous Variables:
   * Gauss or Normal
   * Galton or Log-Normal
   * Exponential
   * Incomplete Gamma
   * Incomplete Beta
   * Weibull

As in Systems Security the interest is generally in the extremities ("tails") of these distributions, that is to say, the higher or lower quantiles, the laws called Extreme Values are also very useful and will be developed in greater detail:
   a) Type I: Gumbell's Law;
   b) Type II: Frechet's Law;
   c) Type III: Weibull's Law.

5.2 Selecting a Law of Probability

The selection of a Probability Law is initially linked to the nature and knowledge of the studied, discrete or continuous phenomenon. In a second step, the choice is guided by the distribution way measured by the asymmetry coefficient and by the flatness coefficient (or kurtosis). In a third step, the choice is guided by the number of parameters that intervene in the mathematical expression of the repartition function. This number usually varies from 1 to 4. Finally, the very value of these parameters can considerably modify the distribution way, as it is the case of the incomplete Beta law and the Weibull's law.

Choosing "a priori" a law of probability, evidently, has a significant impact on the probability value associated with a given quantile. To substantialize this fact and for example, it is estimated the probability of survival of a system beyond 90 hours and 100 hours, knowing that its average is equal to 50 and its standard deviation is equal to 10. The results are presented on the Table 5.2-A.

The analysis of this table shows the major aspect of the laws of extreme values, which generate the highest probabilities of survival. These laws are however exclusively used for estimating extreme events.

Indeed, the knowledge of the average and standard deviation of a law is usually obtained from a sample whose size is not always coherent with the quantiles that serve as basis for the assessment of the probabilities to be compared to the safety targets.

According to the domain covered by the sample, many laws can be applied for modeling, but, for very large or very small quantiles, the deviations among the laws can become very large and the selection must be oriented towards the laws of extreme values which, by construction, allow better estimation from a safety point of view.
The choice itself must take into account the physical laws of the studied phenomenon, which explicitly integrate the set of active factors, whose importance influences the form of probability density and, consequently, on the coefficients that can serve as indicators.

LEI	$P(T{\geq}90h)$	$P(T{\geq}100h)$
Weibull	$4.0.10^{-9}$	$3.5.10^{-16}$
Gauss	$3.2.10^{-5}$	$2.9.10^{-7}$
incomplete Beta	$7.7.10^{-5}$	$9.2.10^{-7}$
incomplete Gamma	$4.5.10^{-4}$	$3.5.10^{-5}$
Galton	$1.1.10^{-3}$	$1.6.10^{-4}$
Gumble	$3.3.10^{-3}$	$9.2.10^{-4}$
Frechet	$6.9.10^{-3}$	$3.2.01^{-3}$

Table 5.2-A) Comparison of Probability Laws

5.3 Extreme Values Laws

5.3.1 Concept

In systems safety analysis, events of probability very close to 1 or very close to 0, corresponding to very large or very small quantiles, are usually considered. In fact they belong to the tails of the probability distribution function of the studied random variable.

In practice, the knowledge of this variable is obtained from a sampling of observed measures. The probability distribution function of the largest or smallest measure of the observed sample is called Extreme Value Law.

5.3.2 Statistics of Order

Let n be a sample of n independent observations $(x_1, \ldots x_n)$, corresponding to n realizations of a random variable X, representing a parameter or a characteristic of a population of infinite size.

If an infinite number of samples of the same dimension n is drawn from this population and if within each sample the observed values are sorted in ascending order, then the values of order i have a statistical distribution, called the distribution of the quantile x_i, which is "a priori" function of the sample size.

This distribution, called empirical repartition function of quantile x_i, is used to fit the sample to a law of theoretical probability [56, 57].

To determine the Law of Probability of a quantile of a given order, let x be a value of the random variable X such that within the sample there are:

    a) (i-1) independent values less than x (event E1);

    b) (n-1) independent values greater than x (event E2);

    c) 1 value x_i between $x \pm dx$, where dx is infinitely small.

From these descriptive considerations, we will try to determine the law of distribution of the quantil of order i belonging to the interval (x-dx/2, x + dx /2). The following assumptions are then made:

    a) the n realizations of x_i independents of a random variable X are equivalent to an x_i realization of n independent random variables X_i;

b) "a priori", the n runs or realizations x_i are equiprobable, that is to say, the repartition function F_i of each of the random variables X_i to the same continuous and uniform law F on [0,1].

For *n* and *i* data, the events E1, E2 and E3 defined above are well identified and their respective probabilities P1, P2 and P3 are easily calculable since the runs are independent:

a) $P1 = \Pr(E1) = \Pr(X < x) = (F(x))^{i-1}$

b) $P2 = \Pr(E2) = \Pr(X > x) = (1-F(x))^{n-1}$

c) $P3 = \Pr(x-dx/2 < X < x+dx/2) = f(x)dx$

It should also be noted that there are:

a) (i-1)! ways to observe E1, corresponding to i-1 permutations of i-1 values less than x;

b) (n-1)! ways to observe E2, corresponding to n-1 possible permutations of n-1 values greater than x;

c) only one (1) way to observe E3.

By sequence, the event (E1, E2, E3) such that E1 is reproduced (i-1)! times, E2 is reproduced (n-1)! times and E3 is reproduced once (1) is distributed according to a multinomial law. If G_i is its repartition function, we have:

$$dG_i = \frac{n}{(i-1)!\,(n-1)!} \cdot [F(x)]^{i-1}\{1-[F(x)]^{n-1}\}\,f(x)\,d(x)$$

We recognize in this expression an incomplete Beta law whose mathematical expression of probability density is written:

$$d\beta\,(a, b, t) = \frac{t^{a-1}\,(1-t)^{b-1}}{\beta(a, b)} \cdot d(t)$$

with a = 1, b = n-1 and t = F(x)

Therefore, the law of repartition of the F_i value of the probability associated with order *i* is written:

$$dG_i = \frac{1}{\beta\,(i,\,n-1+1)}\,[Fi^{\,a-1}\,(1-Fi)^{\,b-1}]\,dF$$

For i = 1, we obtain the law of probability of the smallest (minima) element of an ordered sample within the ascending order X_1 = inf $(x_1, \ldots x_n)$. Being $\beta\,(1,\,n) = 1\,/\,n$, we have a generic law for distribution of minima:

$$G_1\,(x) = \{1 - [1 - F(x)]^n\}$$

For i = n, we obtain the law of probability of the largest (maxima) element of an ordered sample within the ascending order X_1 = sup $(x_1, \ldots x_n)$. Being $\beta\,(n,\,1) = 1\,/\,n$, we have a generic law for distribution of maxima:

$$G_n\,(x) = [F(x)]^n$$

The distribution of the largest quantile (maxima) will be discussed below. It should, however, be remembered that, given the rules of operations with random variables, a minima problem can be converted into a maxima problem through algebraic transformations of the suitable sample data.

5.3.3 Asymptotic Distribution of Maxima

The trivial limits of the generic law of distribution of maxima (largest quantile) are written:
   a)   For F (x) < 1, we have lim $G_n\,(x) = 0$

$$n \rightarrow \infty$$
   b) For F (x) = 1, we have lim $G_n\,(x) = 1$
$$n \rightarrow \infty$$
More precisely, the equation of $G_n\,(x)$ can be written as:
$$G_n\,(x) = (F(x))^n = (1 - (1\,F(x)))^n$$

For n large enough and 1-F(x) small, therefore x large, we obtain:
$$G_n\,(x) = e^{-n\,(1-F(x))}$$

Gumbel has shown [58, 59] that, respecting certain hypothesis, if

$$\lim_{x \to \infty} \frac{d}{dx} \frac{[1 - F(x)]}{F'(x)} = 0,$$

the limit distribution of X_n when n tends to infinity it is written:

$$G_n(x) = \exp\{-\exp[-n\,F'(x_{on}) \cdot (x_n - x_0)]\}, \text{ where}$$
$$x_{on} = F^{-1}(1-1/n) \Leftrightarrow (x_{on}) = 1-1/n$$

We obtain from this mathematical form of Gumbel's distribution:

$$G(x) = \exp\{-\exp[(x - x_{on})/a]\}, \text{ where}$$

x_{on} is the position parameter

$$a = \frac{1}{n\,F'(x_{on})} . \text{ is the scale parameter}$$

It should be noted the arbitrary character of this law in the absence of knowledge of the parent law $F(x)$, which occurs in almost all real cases.

Assuming that the sample source distribution or parent law is normal (Gauss' Law) of μ and σ parameters, the mathematical expression of the order n quantile distribution is written [60]:

$$G(x_n) = \exp\left\{-\exp\left[\frac{(2.\ln n)^{1/2} \cdot (x_n - \mu - \sigma\,(2.\ln n)^{1/2}]}{\sigma}\right]\right\}$$

5.3.4 Types of Asymptotic Laws

As previously seen, the mathematical form of the probability distribution of the largest quantile made the law of distribution of variable X of the Initial population appear. Gnedenko has shown that whatever the initial continuous distribution law in [0; 1], there are only three types of asymptotic laws of the largest (and also the smallest) quantile observed, called extreme values laws[62]. The type I law is said of exponential decay. The type II and III laws are said of algebraic decay.

a) Type I: Gumbel's Law

$$F(x) = \exp\{-\exp[-(x - x_0)/a]\}, a \neq 0$$

b) <u>Type II</u>: Frechet's Law
$F(x) = \exp\{-[(x - x_0)/b]^{-1/a}\}$,
for $x > x_0$, $b > 0$ and $0 < a < 0.5$
that can also be written as:
$F(x) = \exp\{-\exp[(\ln(x - x_0) - \ln b)/a]\}$
If X follows a Frechet's law, then
$\ln(X - x_0)$ follows a Gumbel's law.

c) <u>Type III</u>: Weibull's law
$F(x) = 1 - \exp\{-[(x - x_0)/a]^b\}$,
$a, b > 0$

5.3.5 Applications of Gumbel's Law

The repartition function is written:
$F(x) = \exp(-\exp(-(x - x_0)/a))$
The reduced variable of Gumbel is defined by:
$u = (x - x_0)/a$, where $F(u) = \exp(-\exp(-u))$
The probability density is then written:
$f(u) = F'(u) = \exp(-u) \,^* \exp(-(\exp(-u)))$

The representative curve is an asymmetric bell curve (the asymmetry is due to the signal of parameter a). The curve has its inflection points at $u = \pm 0.96243$.

The main statistical characteristics are presented in Table 4.3.5-A.

Mean	$\mu = x_0 + 0.577216\,a$
Standard deviation	$\sigma = 1.2825\,a$
Mode	$X_0 = x_0$
Median	$X_{0.5} = x_0 + 0.366513\,a$
Asymmetry Coefficient	$\gamma_1 = \pm 1.139$ (sign of a)
Flattening Coefficient	$\gamma_2 = 2.4$

Table 5.3.5-A) Statistics of Gumbel's Law

The probability corresponding to the mean μ is equal to:
    a) 0.570 if a> 0
    b) 0.43 if a <0

The probability corresponding to the mode x_0 is equal to:
    a) 0.368 if a> 0
    b) 0.632 if a <0

Given F (u) = p, the value of x is determined by the inversion of the mathematical form of the repartition function. Then we obtain for the value x_p of the quantile of probability p:
u = - ln (- ln p), where $x_p = x_0 + a$ ln (- ln p)

The quantile u associated with a period of return (or recurrence) T is directly determined directly:
u = - ln (- ln (1 - 1/T)), where
$x_p = x_0 + a$ ln (- ln (1 - 1/T))

For the assessment of parameters from a sample [62] by the maximum likelihood method we obtain:
$$\frac{\sum x_i \exp (x_i / a)}{\sum \exp (x_i / a)} + a = X$$
$X_0 = - a$ (ln (1/n) $\sum$ exp ln (x_i / a))

By the method of moments we obtain:
    a = 0.7797 $\sigma \leftrightarrow \mu = x_0 + 0.577216$ a
    $x_0 = \mu - 0.45 \sigma \leftrightarrow \sigma = 1.28255$ a
    In order to evaluate the confidence intervals, we have that the distance T between x_p and x_p weighted by s_x can be considered as a random variable:
T = $\underline{x_p - x_p} = X - \mu - \sigma \upsilon_p + \upsilon_p$
where $\upsilon_p = -[0.450+0.7797$ ln $(-$ln $(1 - p))]$

This statistic that measures the distance between the quantile of order p and its estimate from a sample of size n was studied by Bernier [63,64]. He proposes two abacuses to estimate the terminals of the 70% and 95% confidence intervals.

For a risk limit α (confidence 1- α), the terminals of the confidence interval of x_p are written, for a number of observations n:

a) $x_p' = x_p + T_2 (p, m, \alpha) s_x$

b) $x_p'' = x_p + T_1 (p, m, \alpha) s_x$

For example, for $p = 10^{-3}$, $n = 120$ and $\alpha = 0.05$ it is possible to determine through the abacuses:

- $T_1 (10^{-3}; 120; 0.05) = 1.25$

- $T_2 (10^{-3}; 120; 0.05) = - 0.95$

The adjustment to Gumbel's law is not made directly from a sample of observations of a characteristic, but from a "synthetic" sample of the maximum values observed over a duration corresponding to the period or time base associated with the occurrence of extreme events considered.

The procedure for obtaining the "synthetic" sample used for the adjustment is performed in four steps:

a) Fixation of time unit: this time unit T serves as basis for defining the ordination relative to the characteristic of the extreme event; it can be the month, year, etc.
b) Segmentation of initial sample: the initial sample is segmented into n periods of T time units.
c) Within each period the maximum value of the observed characteristic is determined.
d) To group the n maximum values as a "synthetic" sample.

The mean and standard deviation are then calculated from the sample to determine the mode x_p and the parameter to be in accordance with the relations previously presented.

The sample values can also be plotted on a Gumbel's paper, checking the suitability of the adjustment. The Kolmogorov-Smirnov adhesion test allows you to consolidate the quality of the adjustment to a risk limit ⬚ (confidence 1 - α).

5. 3.6 Applications of Frechet's Law

The repartition function is written:

$$F(x) = \exp\left[(-(x - x_0) / b)\right]^{-1/a} \text{ or}$$
$$F(x) = \exp\{-\exp\left[(\ln(x - x_0) - \ln b) / a\right]\}$$

for $x > x_0$ and $b > 0$ and $0 < a < 0.5$

The second form is comparable to that one of Gumbel's Law after logarithmic transformation of the variable and taking into account a lower limit x_0. Frechet's reduced Frecht's is defined by:

$$u = (x - x_0) / b$$

The probability density is then written:

$$f(u) = (1/ab) * [(x-x_0)/b]^{-(1/a)-1} \cdot \exp\{[(x-x_0)/b]^{-(1/a)}\}$$

Frechet's law is represented by a family of parameterized curves in a. These curves have the shape of an asymmetrical bell. The main statistical characteristics are presented in table 5.3.6-A.

Mean	$\mu = b\,\Gamma(1 - a) + x_0$
Standard deviation	$\sigma = b\left[\Gamma(1 - 2a) - \Gamma^2(1 - a)\right]^{1/2}$
Mode	$X_0 = b(1 + a)^{-a} + x_0$
Median	$X_{0.5} = b(\ln 2)^{-a} + x_0$

Table 5.3.6-A) Statistics of Frechet's Law

Given $F(u) = p$, the value of x is determined by the inversion of the mathematical form of the repartition function. Then we obtain as the value x_p of the quantile of order p:

$$x_p = x_0 + b\,\exp[-a\ln(-\ln p)]$$

It must be noted the difficulty of choosing between Gumbel's Law and Frechet's Law from generally limited samples. However, it can be stated that Gumbel's Law is the most used one to adjust the maxima sample of a set of observations.

5.3.7 Selecting a Law of Extreme Values

The extreme value distributions laws (or "Weakest-Link distributions") deal with the probability of occurrence $F^*(x^*)$ of maxima or minima of x^* when a large number of independent events are sampled from an initial distribution $F(x)$.

It was seen that the mathematical form of the distribution of maxima (or minima) made the law of repartition of the random variable X of the initial population appear. It is demonstrated that whatever the initial distribution, there are only three types of asymptotic maxima (or minima) laws:

a) Type I (Gumbel)

b) Type II (Frechet)

c) Type III (Weibull)

There is no absolute criterion for selecting a particular type for modeling a given phenomenon. However, by way of guidance the following recommendations can be made [54]:

a) for phenomena with normal initial distribution (Gaussian), such as resistance of materials, flow of floods of rivers, one must use Gumbel's Law;

b) For phenomena with log-normal initial distribution, such as frequency and magnitude of earthquakes, one should use Frechet's Law;

c) For phenomena with Gamma initial distribution, such as wind speed, wave height, one should use Weilbull's Law.

6. Methods of Analysis and Assessment of Systems Safety

Three general types of complementary analysis can be identified for the assessing the safety level of industrial systems:
a) Event Analysis;
b) Zone Analysis; and
c) Time Analysis.

There are also specific methods applied to well defined domains, such as:
a) Dependent and Common Cause Failures;
b) Human Factors;
c) Mechanics (resistance-loading method, probabilistic fracture mechanics); and
d) "Software" development.

6.1 General Types of Analysis

6.1.1 Event Analysis

The time analysis aims to identify and assess the scenarios of events that lead to undesired events.
Four major groups of event analysis methods can be identified:
a) Inductive Statistical Methods;
b) Deductive Statistical Methods;
c) Combined Statistical Methods;
d) Analytical Methods; and
e) Simulation Methods.

6.1.2 Zone Analysis

The zone analysis aims to identify the place of occurrence and the mode of propagation of the undesired event due to geographical implantation of dangerous or particular elements within the physical arrangement of the considered system. Its methods allow us to especially identify the common cause failure modes due to physical implementation of redundancies.

Image 6.1.2-A shows the fundamentals and objectives of the zone analysis for three elements A, B and C that can interface among one another and propagate undesired events.

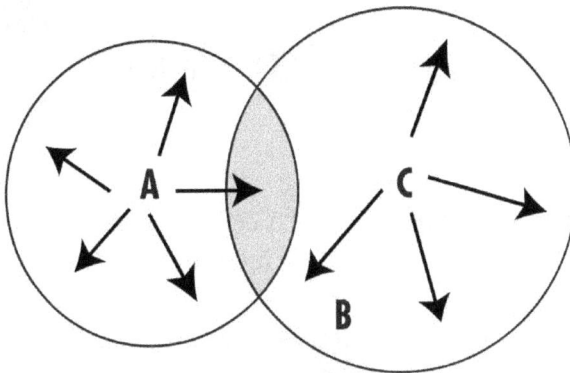

Image 6.1.2-A) Zone Analysis

It is noted in this figure that element B is situated within the field covered by the effects of element C. The elements that are located in the hatched region can then suffer aggressions from the effects of elements A and C.

6.1.3 Time Analysis

The time analysis introduces the dimension of time to the evolution (sequences of events) of dangerous scenarios and delays of risk reduction actions, considering the four times:
  a) t1: information acquisition delay;
  b) t2: information processing delay;
  c) t3: decision delay; and
  d) t4: placing system safety delay;

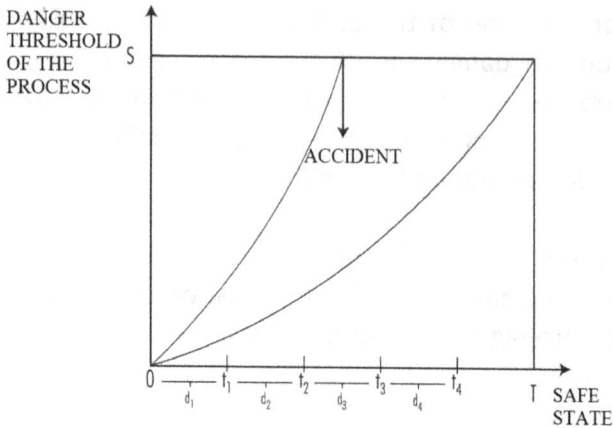

Chart 6.1.3-A) Time Analysis

6.2 Static methods

We distinguish two types of statistical processes within the safety analysis of a system functioning: inductive and deductive.

In the inductive process we start from particular to general. In the light of a system and a failure (or a combination of failures), the effects or consequences of this failure (or a combination of failures) on the system itself and/or its environment are studied in detail. Therefore, for example, the consequences analyses of loss of an airplane's engine or of rupture of a primary circuit pipeline of a nuclear reactor are of an inductive nature.

The main inductive statistical methods are:
   a) Preliminary Risk Analysis (PRA);
   b) Analysis of Failure Modes and their Effects (AFME);
   c) Success Diagram Method (SDM);
   d) Truth Table Method (TTM);
   e) Brief Breakdowns Combination Method (BBCM); and
   f) Consequences Tree Method (CQTM).

In the deductive process, one starts from general to particular. Assuming that the system has a breakdown, the failure modes and causes that lead to breakdown state will be searched. The analysis and investigations that follow the occurrence of catastrophes, to find their causes, are deductive in nature.

The main deductive method is the Tree of Causes Method. It is also possible to consider as a deductive method for safety analysis the so-called "Structured Analysis and Design Technique" (SADT), used in the design of systems for functional analysis and identification of interfaces among component parts.

It is also possible to identify the Causes-Consequences Diagram Method (CCDM), which combines inductive (CQTM) and deductive (CTM) processes.

6.2.1 Preliminary Risk Analysis (PRA)

The preliminary risk analysis was used for the first time in the USA in the early 60's, for safety analysis of liquid propellant missiles [65]. It was then formalized for the aeronautics industry, notably by Boeing [51]. After this use, it has been generalized in numerous industries: chemistry [66], nuclear [1], aeronautics [39] and military [67].

6.2.2 Analysis of Failure Modes and their Effects (AFME)

The method of analysis of failure modes and their effects was used for the first time from the 60's in the field of aeronautics, to analyze the safety of airplanes. Then, the use of this method was generalized in numerous industrial domains:

a) this method is regulatory for safety analysis of aircraft in the USA and France; notably, it was used for the approval of Concorde and Airbus aircrafts [39]; it also served as basis for LEM lunar module's safety analysis;

b) it was recommended for the US nuclear industry after the Three Mile Island accident [4];

c) numerous regulatory texts and standards exist for its purpose, such as: "Military Standards" in the USA [68] and the guide developed by the Institute of Electrical and Electronic Engineers (IEEE) [69];

d) an international standard on method [70] was published by the Commission Electro-technique International (CEI);

e) there is abundant literature about this method and its applications: initially within the traditional domains of aeronautics, space, nuclear, then to chemistry and other domains, such as the automobile more recently.

An extension of AFME is the failure modes analysis, effects and criticality (AFMEC), which considers the probability of occurrence of each failure mode and the gravity class of effects of these failures [71]. So it can be ensured that failure modes with severe effects have very small likelihood to occur.

Another method derived from AFME is the HAZOP (Hazard and Operability) analysis, developed in Great Britain for chemical industry [51].

AFME and AFMEC are preliminary analyses which should generally be supplemented by the use of other methods to identify relevant combinations of failures.

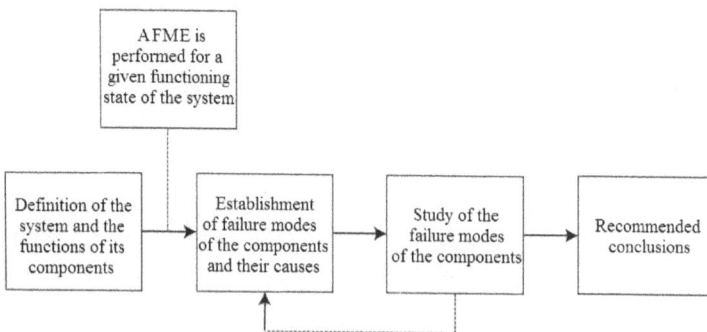

Image 6.2.3-A) Stages of an AFME

The Image 6.2.3-A outlines the steps of preparing an AFME and the Image 6.2.3-B details the main step of establishing failure modes of a component and their causes.

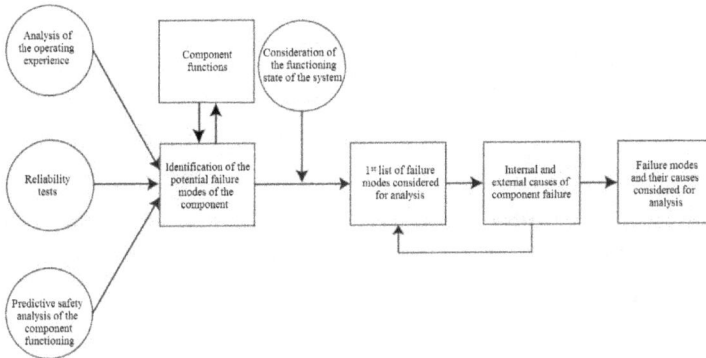

Image 6.2.3-B) Establishment of Failure Modes of Failure of a Component and Its Causes

6.2.3 Success Diagram Method (SDM)

Historically, SDM was the first method to be used to analyze systems and allow reliability calculations. Its origin is confused with the development of mathematical methods associated with reliability [1, 72, 73]. Its limits, however, were then perceived, what gave rise to more refined methods such as TCM and AFME. It remains, however, widely used for relatively simple and non-repairable systems.

From SDM it is possible to deduce the Parts Count Reliability Prediction [74], widely used method for electronic systems, where the failure of a component causes the whole system to fail.

The SDM leads to a modeling of functioning of the system. The success diagram obtained allows the calculation, generally simple, of the reliability (or availability) of the system, supposed to be irreparable.

This method is usable when a detailed analysis of failure causes (and their combinations) is not required, and when there is independence among failures, there is a correspondence between the SDM models and the CTM method models.

6.2.4 Truth Table Method (TTM)

The Truth Table method, whose mathematical aspects are well known in the electro-technical and electronic domains, consists in census of all combinations of functioning states and of breakdown of components, one after another and in studying their effects. This principle, apparently simple, however, turns out to be rapidly unattainable for manual analysis of large systems. However, applications were developed from this method.

The TTM is theoretically the most rigorous method that can be used, the table truth obtained can be the object of a Boolean reduction, obtaining thus pertinent combinations. However, the TTM is no longer applicable when the number of components increases, given the huge number of combinations to consider. However, it is possible to imagine its application to a high level of decomposition of the system, especially at the level of basic functions.

The inconveniences of this method are reduced by its computerization, through the Decision Table Method (DTM) [75], which is derived from it.

6.2.5. Brief Breakdowns Combination Method (BBCM)

The AFME usually only allows to evidence of simple failures and should then be supplemented by the study of failure combinations leading to undesired events. The BBCM is precisely the method that allows us to inductively determine these failure combinations after the FAME is made. The set of abnormal functioning (or failure modes) of a system is thus obtained.

This method was created for the safety analysis of Concorde and Airbus aircrafts [39]. The studies developed in the nuclear area contributed to the development of this method, notably its theoretical foundations [6].

The method is characterized by the introduction of certain specific concepts: internal brief breakdowns, external brief breakdowns, global brief breakdowns. The notion of "effects" allows us to establish the connection among breakdown and the abnormal functioning, undesired events, etc.

6.2.6 Cause Tree Method (CTM)

The Cause Tree Method or Fault Tree was developed in the early 60s to assess and improve the reliability of the Minuteman missile launch system. Eliminating several weaknesses of this project, its use was then considered a great success. In the following years the method was perfected and formalized by Haasl [51].

From 1965 the method was generalized to the most diverse application domains, and it can be said today that it is the most used method for reliability, availability and safety analysis of systems. The method has given rise to a large number of technical publications of which the user guide developed for the USA Authority of Nuclear Safety authority [1] stands out.

The purpose of the CTM is the registration of all causes, of all failures (and their combinations) that lead to occurrence of an undesired event. It allows us to identify the weaknesses of a project, and it is also a way of representing the logic of failures. The existence of numerous calculation software associated with this method favors its use.

However, its use proves itself difficult, if not impossible, for the analysis of elementary systems in interaction and strongly dependent on time. In this case, it may be a useful adjunct to other, more tailored methods. It can also be said that the representation of the method (tree) is an excellent instrument for communication among multidisciplinary teams.

6.2.7 Consequence Tree Method (CQTM)

The Consequence Tree Method or Event Tree was first used between 1972 and 1975 for risk assessment associated with nuclear plants based on PWR and BWR reactors in the USA (Rasmussem Report [41]). It allows us to elaborate and evaluate sequences of events leading to an undesired event called "accidental sequences" and thus characterize the likelihood of the associated consequences.

It has been increasingly used to carry out Probabilistic Safety Analyses of nuclear facilities, being the main recommended method [4].

This method is generally used in connection with the Cause Tree Methods or with the Brief Breakdown Combination Method.

The Image 6.2.7-A schematically presents the steps of elaborating the consequence tree through a deductive procedure.

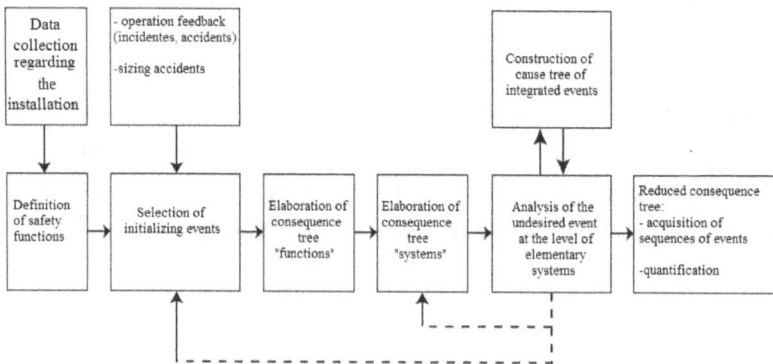

Image 6.2.7-A) Elaboration of a Consequence Tree

Primary pipe rupture	Mission of the reservoir common to 2 systems	Mission of the electric system	Mission of the injection system	Mission of the sprinkler system	Sequences

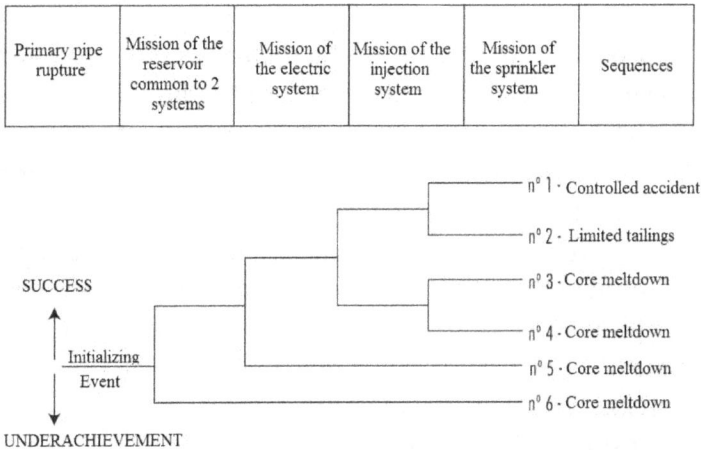

Image 6.2.7-B) Example of a Consequence Tree

6.2.8 Cause-Consequence Diagram Method

The CCDM was initially developed for the reliability and risk analysis of Scandinavian nuclear power plants, by the Risø laboratory, in Denmark, in the early 70s, and now consolidated by the "Risk-Spectrum" computer system, developed by Swedish company RELCON [76].

The simultaneous character of the deductive analysis and the inductive consequence analysis, thus combining CTM and CQTM principles, reveals the method as useful for analysis of specific initializing events, where the order in which the failures occur has a great influence on the occurrence of the undesired event.

6.2.9 Structured Analysis and Design Technique

Analysis by S.A.D.T method [77] is achieved through the construction of a modular, hierarchical and structured model, differentiating the functional aspect (what are the functions to be ensured?) from the material aspect (by what means will the system perform its functions?). The coherence of the various levels of the diagram is carried out through a program. The interest of the method is as follows:

a) to provide a "modeling" of the system as fine as it needs;
b) to show for each function the set of interfaces with the rest of the system, as well as with its environment;
c) to dissociate the functional aspects of design from the material;
d) especially, it stands out for each function or subfunction
e) input data that is transformed by performing the function;
f) outputs corresponding to the transformation of the input data;
g) mechanisms that allow the function to unfold;
h) impositions of the environment.

The application of SADT begins with a more general or more abstract description of the system. This main function contained in a box fragments into more detailed compartments, each representing a larger function of the initial compartment. Each of these compartments decomposes again to expose the information that it contains. The interfaces among various functions are carried out by arrows, defined as follows:

Input	I	Data transformed or used by the function
Output	O	Data created by the function
Mechanism	M	Means that carry out the function
Imposition	C	Impositions of the unfolding function

Table 6.2.9-A: SADT Functions

6.3 Analytical and Simulation Methods

Analytical methods have been developed particularly for reparable systems, modeling the system considered in a state space in which the transaction is made according to a certain probability. The analytical method par excellence is the State Space Method (SSM), initially developed for analysis of Markovian type of processes, and later extended to semi-Markovian and non-Markovian processes [6].

Simulation methods can be interpreted as an extension of analytical methods. They seek the physical modeling of the system's parts, the state transitions of each part and the interface actions among the parts according to the state in which they are. Probabilistic laws of state transitions of the parties, and also eventually of interface actions, are then associated, so that the probabilities of the system's states can be determined by means of random run of Monte Carlo type, and their availability and repairability [78].

The most famous simulation method is the Stochastic Petri Net Method (SPNM) [79].

6.3.1 State Space Method (SSM)

The SSM was developed for the safety analysis of repairable systems functioning [31, 51, 54, 72, 73, 80]. The first industrial use dates back to 1950s and applied to the particular class of stochastic processes called "Markov's processes". Non-Markovian processes were subsequently introduced.

A stochastic process is said to be Markovian, if for each n and $t_1 < t_2$ ... $< t_n$ [11]:

$$P[x(t_n) \geq x_n \mid x(t_{n-1}), \dots , x(t_1)] = P[x(t_n) \geq x_n \mid x(t_{n-1})]$$

Let be considered a system of n components, each having a finite number of functioning and breakdown states. This system is supposed to be repairable, with components repaired after the failure has been detected. It has then [80]:

a) functioning states: are states for which the function of the system is performed, and some of the components may be in breakdown state; the state of good functioning is the one which no component is in breakdown;

b) breakdown states: are states for which the function of the system is no longer performed, with one or more components in breakdown.

The analysis will comprise three parts:

a) census and classification of all states of the system into functioning states or in breakdown states; if each component has two states (functioning and breakdown) and if the system has n components, the maximum number of states is 2^n; over the life of the system, breakdown states may appear as a result of failures or disappear after repairs;

b) census of all possible transitions between these different states and the identification of all causes of these transitions, that are usually due to failures of one or several system components, or to the repair of a component;

c) calculation of probabilities of the system to be found in different states over a period of life, that is to say, the calculation of the safety features of the system.

The qualitative analysis can be represented by a state graph constructed as follows; as shown in Image 5.3. 1-A:

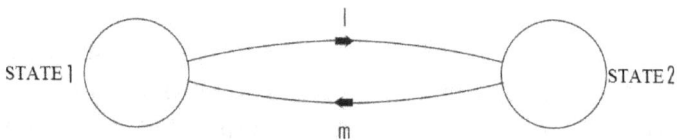

Image 6.3.1-A) Graph of the State of a System With a Component

a) each vertex represents a state of the system;
b) each arc symbolizes a transition between two vertices that unites it;

a transition rate between two states is associated with an arc.

If the probability of passing directly from state i to state j between the instants t and t + dt is $a_{ij}dt$, a_{ij} is the transition rate between states i and j. When transition rates are constant, the process is called Markovian homogeneous. In the system of a component represented by Image 6.3.1-A the transition rate (failure) $1\rightarrow2$ is given by λ, and the transition rate (repair) is given by μ.

A system consisting of two components in active redundancy (Figure 6.3.1-B) has 4 states:

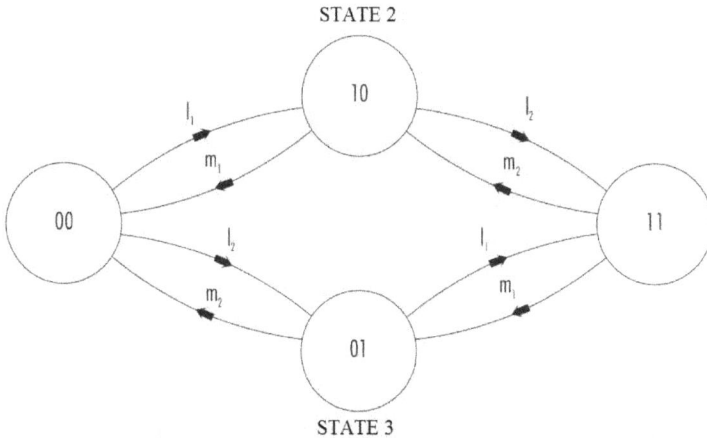

Image 6.3.1-B) Graph of State For a System With Two Components
   a) state 1 (00): two components in operation;
   b) state 2 (10): component 1 in breakdown and component 2 in operation;
   c) state 3 (01): component 1 in operation and component 2 in breakdown; and
   d) state 4 (11) two components in breakdown.
   e) It is observed in this graph that, when the system is in state 4, it can be repaired to return to state 2 or 3. This graph is then adapted to the modeling of the system's availability.

The modeling of system's reliability requires that the probability calculation of the system is in a functioning state (1, 2 or 3), without having been transited by breakdown state [4]. To this end, the breakdown states are rendered "absorbent" by suppressing the transitions to the functioning states. Figure 6.3.1-C shows the graph that models the reliability of the system.

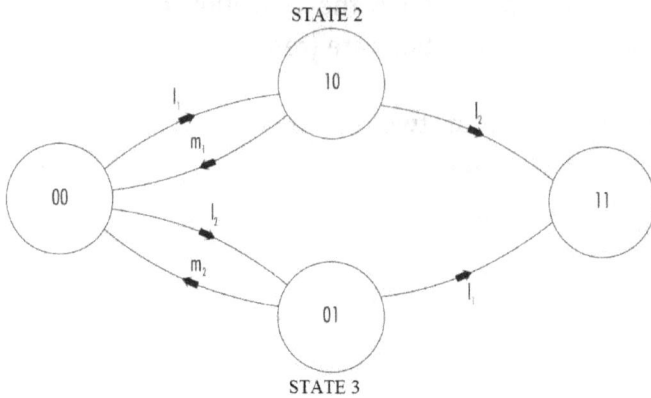

Image 6.3.1-C) Graph of States For Reliability

6.3.2 Stochastic Petri Net

The main advantages of Petri nets compared to Markov's charts are the following [79]:

a) construction of a model by Petri nets is simpler thanks to the possibility of "modeling" conditions and messages that lead to a small number of places and transitions;

b) the obtained results are full of information, especially with regard to the number of transitional runs;

c) they allow us to integrate staggered transitional laws over time.

Petri nets are constructed with the help of "places", "transitions", "conditions" and "messages" [81]:

a) the "places" represent various states of the components of a system (operation, stop, breakdown);

b) the "transitions" represent passages from one state to another and are defined by laws.

Breakdown while functioning	law L	functioning$\rightarrow$breakdown
Breakdown at start	law G	downtime$\rightarrow$breakdown
Repair	law M	breakdown $\rightarrow$ functioning or downtime

Table 6.3.2-A) Transition Laws of Stochastic Petri Nets
   a) the "conditions" are data that authorize or prevent the run of transitions;
   b) the "messages" are the new data after transition runs that will be resumed as "conditions" of other transitions run.

The "conditions" and "messages" make it possible to represent the logic of the functioning of a system. After setting:
   a) duration of functioning; and
   b) a number of "stories" statistically representative,

The functioning of the system can be simulated by the Monte Carlo 78 run. The calculations made allow us then to know:
   a) the "length of stay" in various places, that is to say, availability and (temporary or permanent) unavailability of various components;
   b) the number of runs of "transitions", that is to say, the probability of occurrence of each elemental fact or each feared fact.

6.4 Advantages and Inconveniences of Diverse Methods

6.4.1 Analysis of Failure Mode Their Effects (AFME)

Advantages:
   a) it allows us to systematically identify the effects of failure modes of parts of the system;
   b) it provides a first list of failure modes (and their causes) that compromise operational safety; and

c) it can be extended by a Failure Mode Analysis, their Effects and Criticalities (AMFEC), which allows to classify failure modes according to the couple (probability, gravity).

Inconveniences:

a) the volume of work required to carry it out can be very large, and eventually incompatible with the expected benefits of its results;

b) it only highlights simple failures;

c) it does not lead to a model for quantitative assessment of operational safety.

6.4.2 Success Diagram Method (SDM)

Advantages:

a) the success diagram can be obtained, almost directly, from the functional diagram;

b) combinations of failures that compromise the function of the system are easily identified;

c) the reliability calculation of a non-repairable system is easily performed from the success diagram;

d) an approximate calculation of the availability of a repairable system can be performed when there is independency among the parts of the system under consideration.

Inconveniences:

a) it is not adapted to the analysis of complex relations between effects and causes of failures;

b) it cannot handle repairable systems that have complex maintenance strategies.

6.5.1 Intrinsic Characteristics

Table 6.5.1-A illustrates the comparison among the methods under different aspects of these characteristics.

The classification of the nature of the method is relative to the principles. Actually, in practice, the methods have a mixed nature, for example:

    a) AFME: the research of causes of failure modes is mainly deductive in nature;

CTM: the analyst will not prevent himself from examining the effects of intermediate events and evidenced basic events.

	NAT.			INITIAL EVENTS	Comb ?	REPRESENTATION
	I	D	M			
AFME	Y			component failures modes	N	analysis tables
SDM		Y		system functions	Y	success diagram
TTM	Y			component failures modes	Y	truth table
CTM		Y		undesired event	Y	causes tree
CQTM	Y			initializing event	Y	consequences tree
BBCM	Y			component failures modes	Y	XX
CCDM			Y	initializing event	Y	causes-consequences diagram
SSM	Y			functioning and breakdown states	Y	state graph (Markov)

nat. = nature of method : I = inductive, D = deductive, M = mixed
comb. ? = failure combination research?

Table 6.5.1-A) Comparison of Intrinsic Characteristics of the Methods

As for the ways of presentation associated with the methods, only BBCM has no specific means. However it is important to distinguish the method from its presentation way: so, a cause tree can very well be used as a representation of a brief breakdown obtained by BBCM.

6.5.2 System-Dependent Features

Certain properties of the system to be analyzed help the analyst to choose the method to be used:

a) <u>Non-Repairable System</u>: for such a system, all components are considered irreparable within the limits of a given function; note that a system can be considered irreparable for a given function and repairable for another: in case of an accident A_1, a safety system, that is repairable, must operate for 1 hour; this time is very short to be done any maintenance, which leads to consider it as non-repairable; however, for an accident A_2, must operate for 24 hours, when it is then possible to consider it as repairable. These systems have the simplest analysis, and practically all methods are usable. Note that SSM can also be used, with caution, for partially repairable systems, where certain states do not have transition of repair.

b) <u>Repairable system</u>: in several degrees, the presented methods can treat repairable systems.

- AFME: by identifying failure modes that give rise to a maintenance, it makes a partial judgment of the maintainability of the system;

- DSM: the availability of a repairable system can be calculated when there is independency among parts;

- CTM: can handle repairable systems, even those that are complex; the quantitative analysis suppose independency among parts; to consider complex maintenance strategies, for example, the existence

6.5.2 System-Dependent Features

Certain properties of the system to be analyzed help the analyst to choose the method to be used:

a) <u>Non-Repairable System</u>: for such a system, all components are considered irreparable within the limits of a given function; note that a system can be considered irreparable for a given function and repairable for another: in case of an accident A_1, a safety system, that is repairable, must operate for 1 hour; this time is very short to be done any maintenance, which leads to consider it as non-repairable; however, for an accident A_2, must operate for 24 hours, when it is then possible to consider it as repairable. These systems have the simplest analysis, and practically all methods are usable. Note that SSM can also be used, with caution, for partially repairable systems, where certain states do not have transition of repair.

b) <u>Repairable system</u>: in several degrees, the presented methods can treat repairable systems.

- AFME: by identifying failure modes that give rise to a maintenance, it makes a partial judgment of the maintainability of the system;

- DSM: the availability of a repairable system can be calculated when there is independency among parts;

- CTM: can handle repairable systems, even those that are complex; the quantitative analysis suppose independency among parts; to consider complex maintenance strategies, for example, the existence

6.5.1 Intrinsic Characteristics

Table 6.5.1-A illustrates the comparison among the methods under different aspects of these characteristics.

The classification of the nature of the method is relative to the principles. Actually, in practice, the methods have a mixed nature, for example:

    a) AFME: the research of causes of failure modes is mainly deductive in nature;

CTM: the analyst will not prevent himself from examining the effects of intermediate events and evidenced basic events.

	NAT.			INITIAL EVENTS	Comb ?	REPRESENTATION
	I	D	M			
AFME	Y			component failures modes	N	analysis tables
SDM		Y		system functions	Y	success diagram
TTM	Y			component failures modes	Y	truth table
CTM		Y		undesired event	Y	causes tree
CQTM	Y			initializing event	Y	consequences tree
BBCM	Y			component failures modes	Y	XX
CCDM			Y	initializing event	Y	causes-consequences diagram
SSM	Y			functioning and breakdown states	Y	state graph (Markov)
nat. = nature of method : I = inductive, D = deductive, M = mixed comb. ? = failure combination research?						

Table 6.5.1-A) Comparison of Intrinsic Characteristics of the Methods

As for the ways of presentation associated with the methods, only BBCM has no specific means. However it is important to distinguish the method from its presentation way: so, a cause tree can very well be used as a representation of a brief breakdown obtained by BBCM.

6.4.3 Truth Table Method (TTM)

Advantages:

a) it leads to a systematic state of all failure combinations of parts of the system;

b) a probability calculation is easily performed from the truth table;

c) the Decision Table Method, derived from it, allows to take into account parts of the system that have more than two states and automatically performing an analysis for certain systems.

Inconveniences:

a) it does not apply to systems that have a large number of parts;

b) it is not adapted to the analysis of complex relations between effects and causes of failures;

c) it is not adapted to the analysis of a repairable system.

6.4.4 Brief Breakdowns Combination Method (BBCM)

Advantages:

a) it allows to identify all modes of abnormal functioning and all failure modes of a system;

b) it leads to a detailed analysis of the causes and effects of failures and their combinations;

c) it allows to analyze a set of elementary systems in narrow interaction and to highlight all their interactions.

Disadvantages:
  a) the analysis can be very long and heavy;

  b) it is hardly usable for "very sequential" systems or has complex maintenance strategies;

  c) the quantitative analysis may require the use of other methods.

6.4.5 Consequence Tree Method (CQTM)

Advantages:
  a) it allows to identify all the consequences of an initializing event;

  b) it provides the probability of sequences of unacceptable events; this calculation is facilitated when all events are independent, but otherwise complex (use, for example, of semi-Markovian processes);

  c) it highlights the interactions among elementary systems.

Inconveniences:
  a) it does not allow to demonstrate the completeness of initializing events;

  b) it usually requires the use of other methods for the analysis of generic events;

  c) it does not allow to consider complex maintenance strategies on elementary systems that intervene to control the initializing event.

6.4.6 Cause Tree Method (CTM)

Advantages:

a) it allows to identify all the causes and their combinations of failures leading to a desired event;

b) it shows certain forms of dependency among failures;

c) it provides a list of minimum cuts of an undesired event;

d) it generally allows a calculation of the probability of occurrence of an undesired event;

e) it applies to a wide variety of systems.

Inconveniences:

a) it can lead to causes tree of considerable size and difficult to be handled, even with the help of computer codes;

b) it always allows identifying, for a complex system, sequences of failures that lead to an undesired event;

c) it does not allow you to handle repairable systems that have complex maintenance strategies.

6.4.7 Cause-Consequence Diagram Method (CCDM)

Advantages:

a) it allows to simultaneously identify the causes and consequences of an initializing event;

b) it shows the dependencies between the causes and the consequences of an initializing event;

c) it has, theoretically, the advantages of both CTM and CQTM.

Inconveniences:

    a) the quantitative analysis combines the quantitative analyzes of CTM and CQTM, implying that automatic calculation using computer codes can only be done for particular cause-consequence diagrams;

    b) its use is difficult for a complex set of elementary systems.

6.4.8 State Space Method (SSM)

Advantages:

    a) it allows to identify all the states of functioning and breakdown of a repairable system, as well as all their transitions;

    b) it provides reliability, availability and maintainability measures of a repairable system;

    c) it considers complex maintenance strategies.

Inconveniences:

    a) the quantitative analysis can become very complex, even for a system that has a large number of states;

    b) the quantitative analysis is only relatively easy for Markovian and Semi-markovian processes.

6.5 Comparison of Several Methods

For the presented methods of analysis, the following are distinguished:

    a) intrinsic characteristics, "a priori" independent of those of the system to be analyzed;

    b) dependent characteristics of the system to be analyzed.

of a single repairer for numerous parts, the SSM becomes indispensable;

- CQTM: initializing events and generic events can be associated with repairable systems; however, the independency among these events is admitted as a general rule;

- BBCM, CCDM: the quantitative analysis can take into account the repairable character of the parts;

- SSM: is the method par excellence to handle repairable systems; it becomes irreplaceable for considering complex maintenance strategies.

c) <u>Sequential System</u>: for a sequential system, the success of the mission depends directly on the order of how failures arise; usually, inductive methods are more suited to these systems than deductive methods, and it is all the more remarkable as the complexity of the system increases one; indeed, it becomes difficult to identify the failure causes, for example, using CTM, when causes must produce themselves according to a complex sequence; therefore cause trees have been abandoned, for a long time, for analysis of aircraft crashes and nuclear reactor core meltdown.

d) <u>Multifaceted System</u>: the analysis of such a system is complicated; only more elaborated methods such as CTM, CQTM and SSM treat them under certain conditions.

6.6 Criteria for Selection of Methods

Numerous criteria intervenes in the choice of one or several methods of analysis: they are dictated by technical imperatives or by habit or familiarity of analysts. The criteria are classified according to determining factors:

    a) <u>Criteria related to the targets of the analysis</u>:

reliability, availability, maintainability or security assessment?

- is the undesired event known? Are there other undesired events to be considered?
- quantification is required? Or a qualitative analysis may be enough?

A clear definition of the targets usually leads to limiting the list of possible methods to be used.

    b) <u>Criteria related to the system to be analyzed</u>: the nature of the system and its complexity are important elements while choosing methods; it should be noted that the complexity is not necessarily directly linked to the number of component parts: the system consisting of a large number of parts that can be grouped into macro components may prove to be easy to analyze; what really makes the analysis complicated are factors such as the existence of redundancies, possibilities of human intervention in operation, the repairable character of components, the presence of software. A first functional and technical analysis proves itself to be indispensable before a decision on the selection of methods, examining the properties of the system (repairable, non-reparable, sequential).

    c) <u>Criteria related to analyses already carried out</u>: designers and operators did not wait for safety analyses to predict breakdowns and accidents, and consequently take them into account when designing and operating existing systems; this way, for each nuclear plant, for example, the list of incidents and accidents considered for sizing constitutes a first list of initializing events. In general, all accumulated knowledge in terms of predicting the behavior of the system constitutes an important element in guiding the selection of methods.

d) <u>Criteria related to the resources of analysis</u>: regarding the targets, the methods have different effectiveness and also require different means; it is appropriate to adapt from the outset of the methods to the available resources, in terms of specialists and computer codes, as well as in terms of functioning databases; it is illogical to use sophisticated quantitative methods if reliable data are not available.

6.7 Specific Methods

6.7.1 Dependent Failures Analysis Methods

The main methods of predictive analysis, previously presented, allow us to highlight the dependencies among failures. However, the relative difficulty of predicting all these dependencies leads to multiply the approaches using more specific methods.

The many advantages of these methods, classified according to the level of their effects on a set of systems in interaction are summarized by Table 6.7.1-A, where the sign + gives an approximate and qualitative indication of the interest of each of these methods.

	AFME	CTM	BBCM	CQTM
external initiators				
internal initiators			+	+

elementary systems	AFME	CTM	BBCM	CQTM
functional depedencies			+	+
common equipment dependencies			+	+
physical interactions			+	+
human action dependencies			+	+

components	AFME	CTM	BBCM	CQTM
functional depedencies	+	+	+	+
common equipment dependencies	+	+	+	+
physical interactions	+	+	+	+
human action dependencies		+	+	+

Table 6.7.1-A) Interest of Predictive Analysis Methods For Dependents Failures

Numerous other specific methods have been developed for analysis of dependent failures. They can be grouped into three main families:

a) particular analyses of initializing events of dependent failures;
b) generic causes analyses; and
c) operating experience analysis.

The target of particular studies of initializing events of dependent failures is to analyze in detail the effects of external events (aircraft crash, earthquake, etc.) or internal events (sizing accidents, loss of electrical boards, pipe whipping, etc.) of the system, to limit their consequences, making them acceptable. They allow us to properly dimension elementary systems and affected components.

The analysis of generic causes aims to predict the occurrence and effects of common cause failures caused by one or more generic causes. Among these analyzes, we can cite:

a) predictive analysis of generic causes [82], where the analyst seeks to identify failures on the system guided by the previously presented classification; for potential sources of failures, the analyst checks the constructive dispositions that allow to deal with these failures;

b) zone analysis [39], which aims to analyze the dependencies among failures that result from the "geographic" location of certain components or subsystems; it uses:

- installation procedures;

- physical arrangement of components;

- identification of possible breakdowns; and

- identification of maintenance errors.

c) human factors analysis, which due to its importance is the subject of special methods.

The detailed analysis of the functioning experience is an endless source of dependent failures. It requires standardized and systematic collection of all incidents affecting components and systems. The wealth of this analysis depends very much on the quality of this data collection.

To calculate the probabilities of dependent and common cause failures, two types of methods are identified:

a) explicit methods, based on a precise knowledge of the causes of these failures, that allow the application of the general formula of conditioned probabilities; and

b) parametric methods, that are based on the statistical modeling of failures, without research and enumeration of their causes.

Three parametric methods are identified, that above all deal with common cause failures:

a) parameter β method [83];

b) multiple Greek letters method [84]; and

c) shock method [85].

The results of the analysis of operation experience of nuclear power plant safety systems show the relatively high proportion of these dependent failures dependent on incidents and incidents that actually occurred. They show that the gains in availability or reliability provided by redundancies are less than theoretically predicted. Thus, for the most common components in safety systems, such as pumps, valves, motors, the gain can be estimated from 5 to 20 for an order of redundancy of 2. An increase in redundancy, up to the order of 3 or 4, leads to very limited supplementary gains, at most from 2 to 10. From this point on, gains are marginal [86].

For indicative purposes and without completeness pretension, the following are some means of prevention generally used to reduce the importance of these failures:

a) During the design phase:

- Prevention of initiators that give rise to common cause failures: constructive arrangements to control the effects of:

  -environment and its aggressions;
  -accidental environment of internal source to the installation, the main plausible or hypothetical accidents being considered in the project in a conservative way.

- Prevention of dependencies among elementary systems:

- physical and geographic separation of redundant systems;
- safety functions separation ensured by different systems;
- functional diversity and systems diversity;
- different auxiliary systems;
- possibility of periodic tests;
- optimization of man-machine interface (automation of human actions that cannot be in the required time with sufficient reliability, clear, accurate and proven in simulator operational procedures, consideration of predictable human errors);
- systematic research of dependencies by predictive analysis methods;
- Prevention of dependencies among components:

  -physical and geographic separation of redundant components;
- Diversity of redundant components (designers and different manufacturers);

  -fail safe modes of components;
  -possibility of periodic tests;
  -systematic research of dependencies by predictive analysis methods;

b) During the operation phase:
- Prevention of dependent failures by systematic and detailed analysis of all incidents and accidents occurred on the facility and on similar facilities;

- Prevention of human error:

  -training and motivation of operators, presence of several operators, diagnosis of incidents by two independent and separate teams, using different means;

-interdiction of simultaneous maintenance of redundant components, control by other maintenance teams carried out on important components.

The results of the analysis of operation experience of nuclear power plant safety systems show the relatively high proportion of these dependent failures dependent on incidents and incidents that actually occurred. They show that the gains in availability or reliability provided by redundancies are less than theoretically predicted. Thus, for the most common components in safety systems, such as pumps, valves, motors, the gain can be estimated from 5 to 20 for an order of redundancy of 2. An increase in redundancy, up to the order of 3 or 4, leads to very limited supplementary gains, at most from 2 to 10. From this point on, gains are marginal [86].

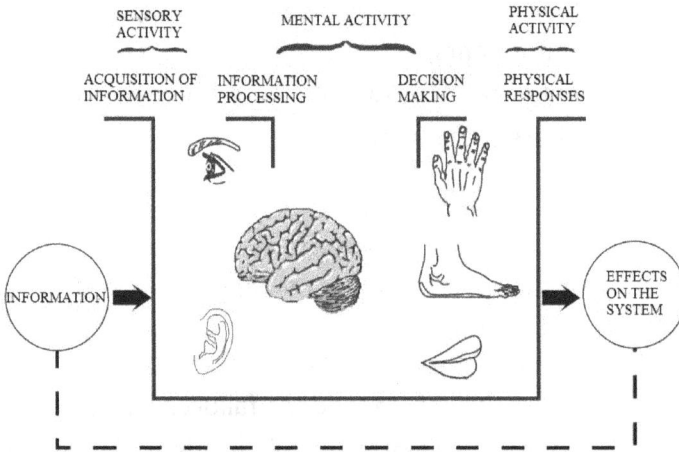

Image 6.7.2-A) Schematic Representation of Behavior of the Human Operator

The most commonly used procedure in the nuclear industry for predictive assessment of human reliability is that of Swain [87]. There are, however, other methods, generally less complete, but better adapted to particular problems.

In order to guide the user, the US Electric Power Research Institute (EPRI) has developed the SHARP (Systematic Human Action Reliability Procedure) [88], which helps in the choice among different methods and the systematic and verifiable conduction of the analysis.

6.7.3 Mechanics of Structures

The development of large mechanical structures, such as pressure vessels of electric-nuclear reactors and ocean oil platforms, generally single or very small in number, whose rupture could lead to very serious consequences and for which over-dimensioning can be demonstrated technically or economically unfeasible, it needs to have methods of predictive assessment of its reliability.

The high quality required for these structures and their small number make it difficult, if not impossible, to use statistical methods. The only way to evaluate their reliability lies in the development of probabilistic models of the mechanical behavior of these structures.

We then identify two probabilistic models that allow us to model the mechanical behavior of the structures, according to the type of charging they are subjected to:

a) Method of Resistance / Charging (R/C), that treats these two variables as random and determines the probability distribution function of the relation R-C (Chart 6.7.3-A), which in negative case implies in the ruin of the structure [89-91];

b) Probabilistic Mechanics of Fracture, that treats the problems associated with the propagation of cracking in components subjected to cyclic charging of fatigue (K_1

coefficient of stress amplification and K_{1C} of tenacity, or resistance to the brutal propagation of a cracking).

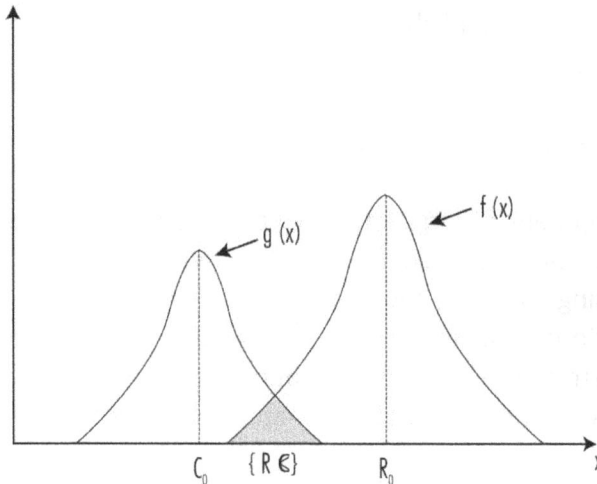

Chart 6.7.3-B) Method Resistance-Charging

6.7.4 "Software" Development

The increasing size of computer codes and their penetration in virtually all industrial sectors make the problem of their reliability crucial. If we consider software of automation and control of industrial facilities and vehicles, to reliability, associated with the success of the mission, adds the security, since a failure can lead to an undesired event with catastrophic consequences.

Such findings have led code designers to develop numerous techniques to obtain the most reliable possible programs, starting from the early stages of their design. In this way, design and development procedures were progressively developed, addressing themes as diverse as languages and programming methods, drafting and validation of functional specifications. These creation methods are, however, considered insufficient, and tests as complete as possible are performed to improve the quality and reliability of the codes.

Despite all these methods, it is difficult, if not impossible, to ensure that a code does not contain any flaws. Parallel to these methods, works have been developed to measure the "confidence" that can be attributed to a program [92].

7. General Procedure of System Safety Analysis

7.1 Concept

The purpose of the procedure is to ensure an appropriate level of safety to the system under analysis, optimizing exclusively made safety efforts with those made for the technical success of the mission.

The procedure concerns the implementation of safety in a new system at the design and development phase. The measures taken, however, cover the operation phase of the system. Preliminary information is:

a) set of accident scenarios identified by the Preliminary Risk Analysis; and

b) likelihood of occurrence of the initializing events of these scenarios.

The expected results correspond to an action program related to each of the stages of the procedure and assessment of the residual risk associated to each one of them.

7.1.1 Description of the Procedure

The safety analysis procedure is divided into four steps, each one corresponding to a set of specific identification and risk reduction actions related to the considered accidental scenario. This, in turn, is divided into a sequence of three events:

a) initializing event (E_0);

b) undesired event (Ei); and

c) accident (Ea).

The position in time of the four steps of the procedure related to the instant of occurrence of the three events corresponds to preventive actions (intrinsic or integrated safety E_1 and implemented safety E_2) or protective actions (E_3 safeguard and E_4 emergency), whose failure probabilities are noted P_1, ..., P_4. The Image 7.1.1-A visualizes this concept. The probability of the accident will then be defined by:

$$P(Ea)= P(Ei)^* \, P(\text{failure } E_1 \,|\, Ei)^* \, P(\text{failure } E_2 \,|\, \text{failure } E_1) \, {}^*P(\text{failure } E_3 \,|\, \text{failure } E_2) \, {}^*P(\text{failure } E_4 \,|\, \text{failure } E_3)$$

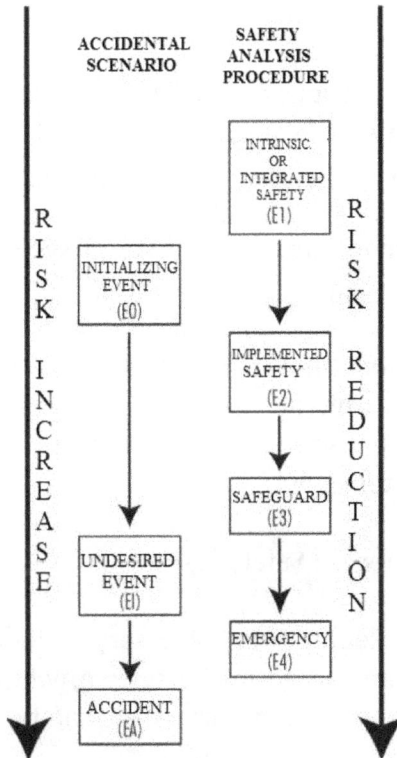

Image 7.1.1-A) Safety Analysis Procedure

7.1.2 Step 1: Intrinsic or Integrated Safety (E_1)

It is related to a "minimum risk design". At this stage, the designer strives to create and implement a configuration for the system that enables the success of its mission. This target has, however, a constant concern to eliminate or decrease "a priori", through an adapted design, the safety measures identified as necessary during the operational life of the system.

This means that, at this stage, no material, logical or human means for exclusively safety purposes should be added. Only material or logical projections of the necessary functions for the technical fulfillment of the mission should be designed as safely as possible regarding the identified risk.

Two possible types of actions respond to this requirement:
a) improvements in global configuration of the system or of an elementary system or associated subsystem; and

b) improvements to components or system items through their selection.

However, these actions should not have a negative impact on the probability of success of the mission, because at this stage there should be no compromise between the success of the mission and the safety of the system.

7.1.3 Step 2: Implemented Safety (E_2)

At this stage, safety elements not necessary to the technical success of the mission are implanted in the system, which in some cases implies a compromise between success and safety.

These elements integrated into the system configuration to eliminate or reduce residual risk constitute "safety or in depth defense barriers". They are material, logical or procedural devices that prevent or delay the evolution of a scenario, in order to counteract the occurrence of the undesired event.

To justify its implementation in the system, the barriers must meet three basic principles, that are closely connected to the principles of probabilistic language use in SoF:

a) 1st Principle: the barriers must be reliable, that is to say:

- effectively adapted to cover the undesired event whose occurrence is to be avoided;
- robust, that is to say, sufficiently sized regarding the undesired event to avoid its premature failure;
- durable, that is to say, effective over time.

b) 2nd Principle: if several barriers of the same reliability are implanted in redundancy, they, as well as the element of reconfiguration, if it exists, should be:

- separately tested;
- independent with regard to common cause failure modes; and
- impermeable to all processes of failure propagation or its effects, that is to say, the failure of one of the barriers should not imply the failure of others and/or the element of reconfiguration, if it exists.

c) 3rd Principle: the effectiveness or reliability of the proposed barriers must be systematically justified by safety analysis and validated by adapted experiments.

7.1.4 Step 3: Safeguard (E_3)

This stage corresponds to the implementation of a set of emergency actions that follow one or more confirmed alarms, announcing the occurrence of the undesired event and whose direct causes must be controlled, that is to say, permanently neutralized or delayed to allow the system to be returned to a safe state.

This operational safety stage groups the set of preventive measures with regard to the undesired event. It shows the need to predict still in the design phase:

a) tests and alarms associated to elements whose failure is critical, which implies in the implantation of sensors and monitoring channels perfectly covering the phenomenon under surveillance; and

b) an effective strategy of driving the system or its environment to a safe state.

7.1.5 Step 4: Emergency (E_4)

This stage corresponds to a set of emergency actions whose objective is to mitigate the consequences of the accident, that is to say, to minimize the gravity of the potential effects of the undesired event, after its occurrence. This step groups the set of protective measures with regard to the undesired event.

The system design logic for each of these operational phases and for each identified event then consists of predicting a set of specific and effective actions that decrease the gravity of direct and indirect consequences of each of these undesired events.

7.1.6 Simplified Application Example

Let be a closed location where specific operations are carried out on a vessel containing gas at high pressure, according to the diagram shown in Image 7.1.6-A.

OPERATIONS IN CLOSED LOCATION

HIGH PRESSURE GAS VESSEL

Image 7.1.6-A) Application Example

The accident Ea corresponds to the death of at least one operator due to the fragments of the reservoir explosion. Examples of actions corresponding to the four stages of safety procedure would be:

a) E_1:

Design, manufacture, assemble, commission and periodically inspect the vessel according to standards and quality assurance procedures adapted;

b) E_2:

-install pressure relief valves;
-place physical barriers around the vessel;

c) E_3:
- install an overpressure monitoring and control system to alert operators in case of imminent risk of explosion (exceeding a critical pressure limit);

d) E_4:
-provide emergency exits of the place;
-provide aid team;
-provide bed in the nearest hospital.

7.2 Failure Modes Analysis

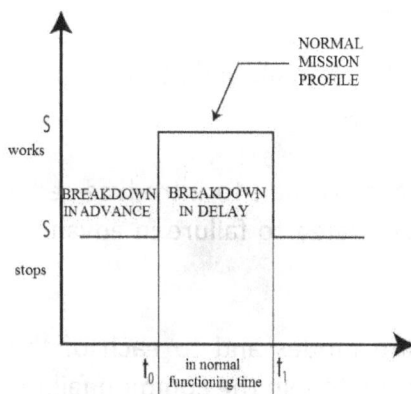

Chart 7.2-A) Failure in Advance and Failure in Delay

Let be a System S whose mission, which requires its nominal functioning between the instants t_0 and t_1, is represented by the Chart 1.7.2-A.

Four generic failure modes are considered in analyses by statistical methods of AFME type:

    a) unexpected functioning or failure in advance, which corresponds to the start and functioning of S before t_0;

    b) unavailability at start, which corresponds to the non-functioning of S at t_0;

    c) premature stop or failure in delay or even unexpected non-functioning of S before t_1; and

    d) stop failure, which corresponds to the functioning of S after t_1.

These generic failure modes are also combined with failure modes derived from unacceptable variations of the functioning parameters of S between t_0 and t_1.

In practice, it is generally considered:

    a) failure in advance;

    b) failure in delay, usually linked to unavailability at the time of departure.

The action of the designer is then to introduce redundancies that meet the safety targets related to failure in advance and failure in delay.

To these intrinsic failure modes and to each of the elements in redundancy, it is necessary to join the common failure mode, which implies, under certain circumstances, the simultaneous inefficiency of redundancies.

7.2.1 Failure in Delay and Failure in Advance of Elements in Total Redundancy

Let the functional scheme of S be presented by Image 6.2.1-A, where elements A and B are independent "a priori" and connected in series.

The reliability diagrams of S for failure in delay and for failure in advance are shown respectively by Images 6.2.1-B and 6.2.1-C.

For S to suffer failure in delay in S_{de}, it is necessary that A or B suffer failure in delay, that is to say, directly applying Poincaré's Formula:

$$S_{de} = A_{de} \cup B_{de}$$
$$P(S_{de}) = P\,(A_{de} \cup B_{de}) = 1 - (1 - P(A_{de}))^{*}(1 - P(B_{de}))$$

In order for S to suffer a failure in advance S_{av}, it is necessary that A and B suffer a failure in delay, that is to say, by directly applying Poincaré's Formula:

$$S_{av} = A_{av} \cap B_{av}$$
$$P(S_{av}) = P\,(A_{av} \cap B_{av}) = 1 - (1 - P(A_{av}))^{*}(1 - P(B_{av}))$$

The reliability diagram for failure in delay always has the same form as the functional scheme. The reliability diagram for failure in advance always has the form of the dual functional scheme.

Image 7.2.1-A) Functional Scheme

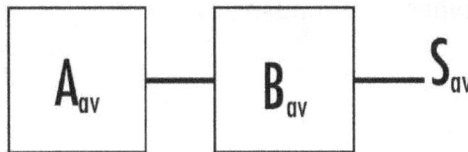

Image 7.2.1-B) Failure in Delay

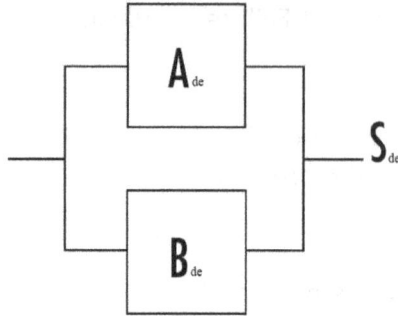

Image 7.2.1-C) Failure in Advance

The Table 7.2.1-D presents the expressions of failure in delay and failure in advance for systems with simple redundancies of identical elements.

TYPE	FAILURE ADVANCE	FAILURE DELAY
simple element	P_{av}	P_{de}
parallel	$P_{av}(2-P_{av})$	P_{de}^2
series	P_{av}^2	$P_{de}(2-P_{de})$
series/parallel	$P_{av}^2(2-P_{av}^2)$	$P_{de}^2(2-P_{de})^2$
parallel/series	$P_{av}^2(2-P_{av})^2$	$P_{de}^2(2-P_{de}^2)$
series/half parallel	$P_{av}^2(2-P_{av})$	$P_{de}(1+P_{de}-P_{de}^2)$
parallel/half series	$P_{av}(1+P_{av}-P_{av}^2)$	$P_{de}^2(2-P_{de})$

Table 7.2.1-D) Expressions for Failure in Advance and Failure in Delay of Simple Redundancies

The Chart 7.2.1-E, on the next page, represents the relative positions of the probabilities of failure in advance and failure in delay for these types of redundancies.

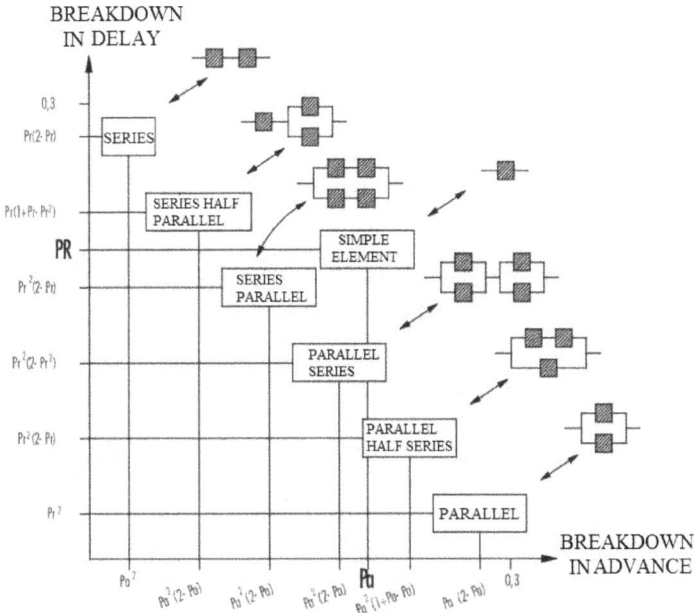

Chart 7.2.1-E) Comparison of Probabilities For Failure in Advance and Failure in Delay of Simple Redundancies

7.2.2 Failure in Delay of Elements in Partial Redundancy

A system in partial redundancy (k / n) is formed by n elements and suffers a failure if at least a number k of elements fails (k <n). To solve this case, the binomial law is used.

The probability that at least k elements fail is given by:

$$P(S)=P(K\geq k,n)=\sum_{i=k,n}\binom{n}{i}p^i(1-p)^{n-i}$$

$$\binom{n}{i}=\frac{n!}{i!(n-1)!}$$

K = number of elements in failure
P = probability of failure of an element

The reliability is expressed by the relation:

$$R(t)=P(S)=P(K<k-1,n)= \sum_{i=0,k-1} \binom{n}{i} p^i(1-p)^{n-i}$$

For example, for n = 4 and k = 3 (redundancy 3/4), we will have:
$$P(S)=4p^3(1-p)+p^4$$

7.2.3 Common Cause Failure Modes

Regardless of the random and individual failure of the elements of a redundancy, the redundant element type, the source of the inputs in the element (power supply, control and command signals) and its physical implementation in the system must be considered in the design:

a) in the first case, design errors, fabrication or maintenance of identical elements can cause simultaneous common cause failure modes;

b) in the second case, external failures to the elements, such as loss of power supplies, can cause simultaneous unavailability of elements; and

In the third case, an external aggression can hit simultaneously or propagate through the elements, in case there is no adequate physical security.

These aspects, if not properly checked, may compromise the effectiveness of redundancies.

It is then necessary to differentiate, in the assessment of the reliability of a redundant system:

a) independent failure rate, individual of each of the elements, noted as λ_i, that integrates the failure rates in advance and in delay; and

b) common cause failure rate λ_c.

Considering the two types of incompatible failures by definition it is possible to as the global failure rate λ of each element and the part β due to common mode failures:

$$\lambda = \lambda_i + \lambda_c$$
$$\beta = \lambda_c / \lambda = \lambda_c / (\lambda_i + \lambda_c)$$
$$\lambda_i = (1-\beta)\lambda$$
$$\lambda_c = \beta\lambda$$

Considering a system S with total parallel redundancy of elements A and B, with $\lambda_{iA} = \lambda_{iB} = (1-\lambda)\beta$ and $\lambda_c = \beta\lambda$, three possibilities can be considered, for which the probability can be calculated:

a) two elements work - this hypothesis corresponds to the classic calculation of the failure rate of a total redundancy in parallel:

b) $P(S_0) = P(A \cap B) = (1-\lambda)(1-\lambda) = (1-\lambda)^2 =$
= $1-2\lambda+\lambda^2$, if λ is small, $(1-\lambda)^2 \approx 1-2\lambda$

c) one of the elements works: this hypothesis corresponds to the case which A fails ($\overline{A}$) and B works, or B fails ($\overline{B}$) and A works; this implies the non-existence of failure in common cause mode:

$$P(S_1) = P(\overline{A}, B) \cup P(A, \overline{B}) =$$
$$= P(\overline{A} \cap B) + P(A \cap \overline{B}) =$$
$$= P(B)* P(\overline{A} \mid B) + P(A)* P(\overline{B} \mid A)$$

Being A and B independents,
$$P(\overline{A} \mid B) = P(\overline{A}) \text{ and } P(\overline{B} \mid A) = P(\overline{B})$$
$$P(S_1) = 2(1-\lambda)(1-\beta)\lambda$$

d) Two elements fail - this hypothesis implies that A and B suffer independent failure (I) or A and B suffer a common cause failure (C):

$$P(S_2) = P(A \cap B)_I + P(A \cap B)_C =$$
$$= \lambda\beta + (1-\beta)^2 \lambda^2$$

7.3 Probability Assessments from a Law of Mortality

Let T be the random variable representing the life span of a system and F its probability repartition function. If $t_2 > t_1$, the conditional probability of the event $\{T > t_2 \mid T > t_1\}$ is given by:

$$P(T \geq t_2 \mid T \geq t_1) = \frac{P(T \geq t_2)}{P(T \geq t_2)} = \frac{1 - F(t_2)}{1 - F(t_2)}$$

This relation allows estimating the probability of survival at time t_2 of a system knowing that it worked correctly until t_1. For example, if the law of mortality of the system is exponential with failure rate λ, the conditional probability is given by:

$$P(T \geq t_2 \mid T \geq t_1) = \frac{e^{-\lambda t_2}}{e^{-\lambda t_2}} = e^{-\lambda(t_2 - t_1)}$$

Note also that the probability of survival (or mortality) only depends, in this case, on the time interval (t_2, t_1), since the failure rate is constant.

7.4 Limitations of Analysis

Here will be presented some important aspects or problems posed by safety assessment and that generally lead to their limits.

7.4.1 Limits of Qualitative Assessment
   a) It is not possible to predict everything! Although evident, this deserves to be remembered to analysts that are carried away by the analytical power of methods that highlight an

impressive number of minimal cuts of an undesired event. Confrontation with the feedback of operation leads, on the one hand, to rely on these methods and, on the other hand, to modesty, especially in the domain of common cause failure modes and of human errors.

b) Conservatism of assessments: the analyst, confronting multiple questions about the effects of component failures will be certain times taken, in lack of a functioning analysis concluding otherwise, to assume that they lead to the undesired event. In general, operational safety assessment requires a very detailed knowledge of the system and the effects of minor failures of its components that is not available during dimensioning. In fact, it is generally sufficient that the designer demonstrate, in order to justify the dimensioning, that there are margins with regard to criteria, such as accidental postulated situations. However, the safety operational specialist would like to meet and assess theses margins. It results then that the safety operational models are generally conservative, at least regarding identified failures.

c) Dependency among failures: their identification is critical, because it can deeply modify the findings of the safety assessment. Particular attention is needed with regard to common cause failure modes and to human errors. Bear in mind that is generally assumed that there are no design errors, which is almost never true.

7.4.2 Limits of Quantitative Assessment

a) Operational Safety Data: quantitative assessment is highly dependent on the quality of reliability, availability and maintainability data; sensitivity studies can compensate for a not very good quality of these data. Generally, the data are relative to similar equipment, its consideration for assessing an ongoing project implies assuming that the new equipment will have at least the same behavior to those that generated the database. There are then two implicit hypotheses:

- quality assurance, qualification and commissioning tests will have a less equivalent level at least equivalent;
- equipment will be operated (periodic tests, preventive maintenance, predictive maintenance) under similar conditions.

b) The existence of uncertainty about these data results both from the random nature of failures and from the imprecise knowledge of environmental conditions of operating components. One of the advantages of safety analysis is to highlight and consider these uncertainties, so it is an obligation to precisely state the limits of the knowledge and knowhow used in the system.

The existence of data, even with large inaccuracies, translates knowledge progress and is preferable to the traditional "engineering judgment" used to impose an opinion without much foundation. However, overconfidence in numbers must be avoided.

c) Operational safety measures: these measures must always be associated with respective uncertainties, especially for decision making analyses. The calculation of predicted values is usually done in an approximate way, which does not make a real deficiency as long as obtaining orders of magnitude is sufficient to reach an analysis target. The permanent and detailed confrontation of the analysis model with the feedback makes it possible to decrease these limits.

7.5 Analyses Validation

Is there a good agreement between the predictive measures of operational safety and the operational measures resulting from the feedback analysis? This fundamental question sets the problem of validating a predictive assessment.

Numerous are the examples of relevant failures or breakdowns identified during safety analysis and that have occurred effectively throughout the life of the system. A first comparison regarding the analog and digital telecommunications cards used by France Telecom [93] shows that the "validation ratio" between the operational value (Vo) and the predicted value (Vp) for the failure rate of these components is, in 75% of the cases, less than two.

A second comparison, regarding 130 systems of quite different nature of the chemical and nuclear industries, carried out by the British National Center of Systems [93], shows that 63% of the predicted values are within the interval $[V_0/2, 2V_0]$ and that 93% are within the interval $[V_0/4, 4V_0]$.

A third comparison refers to the elementary system protecting the turbine of 900MW French nuclear power plants [93], of which there are about 30 in operation. This system causes a sudden shutdown of the plant, in a ratio of 0.5 events/year/reactor, due to system components failures. The predicted value for this number of events is about 1/year/ reactor. The validation ratio is then 0.5.

It is possible to affirm that the studies confirm that the quantitative analysis can predict, with a reasonable error margin (of 2 to 4), the reliability of a system. However, this is not so true for cases of highly redundant systems, where the part of failures due to common causes or human errors can be important or even dominant.

With regard to catastrophic risks, it is difficult to make such comparisons, since they rarely happen. It may be considered, by ways of gross comparison, the occurrence of nuclear core meltdown accidents in electronuclear reactors.

The annual probability of occurrence of this event is 10^{-3} to 5.10^{-5}/year, according to the results of Probabilistic Safety Analyses made for some of these reactors [94]. Two accidents of this type effectively occurred (Three Mile Island, in 1979, and Chernobyl, in

1986). Until 1986, the international experience accumulated for this reactors reached around 3,800 reactors per year, which implies $V_O = 5.10^{-4}$. This value does not seem to indicate a contradiction regarding predictive assessments.

It can be considered in greater detail the comparison between the Probabilistic Safety Analysis, from 1975, about the power plant of Surry, USA, and the Three Mile Island Station accident scenario, since this last one had a design quite similar to the first one. The accidental sequence occurred had been identified by the analysis, as an initiator event "small break in the primary circuit", followed by failure in the cooling safety injection system. The probability of an operator diagnosis error leading to the shutdown of the safety system, that effectively did not occur, was not considered important. A series of precursor events occurred at similar facilities prior to the accident: the lack of a feedback analysis collection system prevented the sequence from being reevaluated. If it had occurred, the importance, under the probabilistic plan, of the sequence would have been revealed and possibly corrective actions would have been taken and the accident might not have happened.

The validation of the operational safety model should be a permanent concern, and must be reviewed based on the feedback.

Predictive safety analyses can only be made and continue using a certain number of hypotheses regarding failure effects, system limits, reparability of components, human factors impact and component failure rates. These hypotheses are likely to evolve throughout the system operation and the analyses should be periodically updated to cover these evolutions and modifications.

The validation of the model by operation experience is therefore an indispensable guarantee of its quality and its coherence with the reality of operation.

7.6 Organization and Management of Safety Analysis

The effective utility of the predictive assessment will heavily depend on its organization and management. Following are some principles that favor the use of methods:

a) The Operational Safety is a specialization: the use of the methods must be done by specialists in OS, because the experts who design and calculate the functioning of a system have difficulties to imagine failures (this cannot happen) and will tend to consider failure as an "aberration", because all efforts are oriented to the success of the mission, making it hard to live, in the same professional, two approaches; specialists in operational safety must necessarily have technological curiosity, ability to learn new technologies, ability to abstract and good morale to live in a world of failures and catastrophes. Experience is indispensable in this field to avoid two extreme and undesirable positions:

- the assessment is useless or unreliable;

- the numbers are unquestionable and decisions should be taken based on them;

- finally, the intellectual honesty of the specialists is indispensable in the field where the engineering judgment has great intervention and where the data can be manipulated.

b) Operational safety is multidisciplinary: predictive assessment has the need to integrate as many factors as possible throughout the life cycle of the system; this ambition can lead experts to a high level of details and to devote to diverse problems such as the behavior of a material in an accidental situation, written form of a procedure, existence and effectiveness of commissioning tests, periodicity of periodic tests; by essence, the assessment of operational safety is then a multidisciplinary activity; the experts should work closely with the designers, operators, specialists in functioning, ergonomics; the assessment of operational safety can prove to be an activity that structures many associated domains: new working and communication relationships within the company are created and new concerns arise; in general, it can contribute, on the one hand, to spread greater rigor and new methods in management, and, on the other, to develop a true safety culture.

c) <u>Operational safety requires organization</u>: considering previous aspects, every important program implies implementation of a true organization of assessments within the company; the potential importance of the impact of such studies requires that the organization adopted, which should be adapted to the problem to be addressed, takes into account the evolution of the company in terms of operational safety and is accepted by those will use the lessons learned from the studies; it is important to highlight the fundamental dissemination of the most relevant lessons to the decision-making levels.

d) <u>Quality Assurance of Analyses</u>: quantity, diversity, potential repercussions of hypotheses manipulated during a safety analysis requires implementation of a true quality assurance of studies; the verification and validation of the qualitative analyses by several specialists are generally wealth in teaching; a detailed analysis of the quantitative results, its systematic verification of values obtained by approximate calculations and/or by several methods are elementary precautions that cannot be forgotten; note that in the field of nuclear safety, an independent control is general requirement.

e) <u>Management of Safety Analysis</u>: usually, the predictive assessment is only valid for a precise instant of the life of the system; the evolution of the project and after the operation lead to programming its review; this can rebound on the methods and encourage the use of software to make these reviews easier.

7.7 Use of Safety Analysis

The Operational Safety is supposed to be the "Science of Failures", that is to say, its analysis, assessment, measurement and prediction. Its quantification targets motivated a considerable advance in the development of qualitative methods that however did not definitely solve the precision problems of the obtained measurements.

However, as it was seen, since the OS approaches knowledge in the Uncertainty Domain, this quantitative precision will always be a utopia. The OS is then much more an intellectual attitude to approach the phenomena of failure than a precise scientific tool. In this respect, the OS has become a solid and inescapable industrial reality.

Since the 80s, quantitative predictive assessments have had increasing repercussions on the design and operation of technological systems. The quantifiable magnitudes that emerge from these analyses aid technical decision-making and are considered as fundamental elements for the specification of a system, as well as functional performance measures and cost.

7.7.1 Use in Design of Systems

a) <u>Comparative Assessment of Different Design Alternatives</u>: when the designer hesitates among several possible configurations for a system, a comparative assessment of operational safety becomes an important aid to the decision; the quantified assessment can be coupled to an economic comparison of the cost of designs; often, the comparison is very significant, because it does not depend very much on the operational safety data adopted; however, this is not systematically true and one must be regarding comparative assessments that only depend on an uncertain quantitative figure.

b) <u>Assessment of Operational Safety of a Project</u>: clearly, methods used for this assessment can evolve over the course of the project, as it goes into detailed; it becomes important to start the first predictive assessments as soon as possible, notably to verify if the great technical options adopted are compatible with the general safety targets; afterwards, it is necessary to check in detail if the targets continue to be respected in the advanced stages of the project; in fact, it is important to remember that a single failure, not previously predicted and whose probability is judged high, can compromise the best elaborated projects, especially when aspects of social perception of risk are at stake.

Usually the prediction is made for the useful life of the system: so it does not seek to predict the probability of premature failures and therefore the expected reliability should be compared to the operational reliability, after eliminating premature failures, that is to say, after performing commissioning tests.

It is important to remember that the safety data of functioning components generally corresponds to the useful life. Then the prediction will be all the more accurate as the qualification and commissioning tests have been correctly specified and performed, this way, eliminating defects of difficult quantification.

This fact addresses the relation between the predictive safety assessment and the quality assurance. The better the quality assurance is, the better the prediction will be. The predictive assessment and the acquisition of good predicted values of operational safety do not exempt, under any hypothesis, quality assurance, on the contrary, makes its reinforcement necessary. Predictive safety assessment may help us rank the priorities of quality assurance priorities and identify the components that should, mandatorily, be subjected to particularly careful procedures.

The performance of predictive assessment of operational safety goes through the elaboration of a real Operational Safety Program, which must have a structure of organization, responsibilities, procedures, activities, abilities and means that compete to the guarantee that the system will meet the relative safety targets to a particular project.

It is equally necessary to implement an operational safety assurance policy to give credibility when achieving the required objectives and to ensure that the operational safety is what it was supposed to be. As components of this policy, we can cite:
   a) operational safety audits;
   b) design reviews;
   c) safety data collection system and subsequent treatment and use; and
   d) continuous surveillance of measures taken for safe functioning.
This warranty policy initiated during designing must, of course, continue during the operation.

7.7.2 "Deterministic" Design and "Probabilistic" Design

Every project of a system must take into account the uncertainty domain of knowledge that being used and the environment where it will be build and operated. Notably, the internal failures of the system and its operating conditions are random. In the light of this problem, two types of design methods can be distinguished: "deterministic" and "probabilistic" design.

a) "Deterministic" Design: schematically, consists in choosing, among a set of plausible scenarios, a reference scenario or envelope, such that if the system is, for example, reliable and safe within the reference scenario, it is equally considered reliable and safe for a "large majority" of scenarios considered less severe; this method then replaces the space of probabilistic scenarios for an estimated equivalent deterministic scenario; this method is systematically based on standards and guides for design, manufacturing and operation developed based on the experience accumulated in many technological domains involved.

b) "Probabilistic" Design: schematically, this method rests on a predictive assessment of operational safety, consisting of identifying the set of plausible scenarios and then associating each of them with a probability and gravity of consequences, that is to say, a risk; by means of a decision criterion associated to the general safety targets of the system, combining both dimensions of risk, the method proposes to define the "great" configuration with regard to the set of scenarios adopted.

It is important to highlight that every design is implicitly probabilistic: for example, there is only one spare tire in passenger cars, which implies to assume that the risk associated with the concomitant occurrence of two flat tires is considered acceptable. However, the name "Probabilistic Design" is reserved for the one developed based on a predictive assessment associated to the fulfillment of quantified safety targets, in other words, explicit.

The deterministic and probabilistic design methods are increasingly used together: to the relative "simplicity" of the deterministic design, the probabilistic design adds the research, acquisition and assurance of an acceptable level of operational safety.

This coupling is particularly remarkable in the case of nuclear power plants, where the deterministic design, based on the "Basic Design Accident" (guillotine rupture of the main reactor cooling pipe associated with the occurrence of the maximum earthquake plausible for the site of the plant), is subjected to a Probabilistic Safety Analysis, which identifies and quantifies the probability and gravity of other scenarios and accidental sequences with more or less severe consequences. This analysis makes it possible to implement improvements to designs, identifying possible "weaknesses" regarding the safety and evaluating compliance with general safety targets, which at the present time do not have regulatory impositions.

The quantified safety targets are currently increasingly used, if not as regulatory imposition, at least as contractual supply imposition. The same is observed in problems of reliability, availability and maintainability, whose quantified targets tend to be generalized in order to, on the one hand, help the designer to rationalize his design options, and, on the other hand, to guarantee the user a minimum acceptable operational safety.

These quantified targets appear more and more in supply contracts as reliability clauses. The methods of predictive and verification target assessment are usually foreseen, associated with penalties when they are not respected. The existence of penalties leads those responsible to pay close attention to the content of these clauses.

7.7.3 Use in Operation of Systems

a) Assessment of a system in operation: although an analysis of the operating experience (incidents, accidents) of an industrial system can give important elements of appreciation, predictive assessments for actual systems in operation become necessary in certain situations; it is the case, for example, of dangerous facilities when one wants to know the probability of occurrence of an undesired event identified during the operation phase; the analysis methods do not differ from those used to assess a project: the operational safety data will, however, be specific to the facility and it will normally be possible to consider the actual operating conditions.

b) Feedback Analysis: imagine an industrial system that has safety problems; the feedback analysis can be approached in two different cases:

- there is no predictive model of accident risk: certain incidents occurred have cast doubt on the safety of the facility, without making it possible to easily assess the existing margins with regard to a major accident; the incident is then considered as an initializing event of or as an event that compromises the operation of the facility after the occurrence of another initializing event; the accidental sequences can be prepared in a predictive way, for example, with the help of TCM: the calculated accident probabilities explicitly explain the existing margins and make it possible to assess the likelihood of a modification of the facility design; such an approach makes it possible to classify incidents according to the gravity of their consequences and, in particular, to identify precursor events;

- there is a predictive model of accident risk: the feedback analysis is introduced in the previous model; when the introduced event (new failure mode, failure mode with probability higher than initially estimated) implies a significant increase in the quantitative accident risk indicator, the event will be considered as a precursor event, which may require a modification in the facility to avoid its repetition or limit its consequences. In general, the predictive assessment makes it possible to rank the incidents, so the most dangerous ones are priority.

c) Definition of periodic tests: generally, the safety systems of industrial facilities do not operate under normal conditions, only being activated when incidents occur; this makes it imperative to test them regularly; the availability assessment of these systems, providing knowledge of operational safety parameters, makes it possible to determine an optimized periodicity for these tests, making minimum the periods of unavailability of the system.

d) Definition of operating procedures: complex industrial systems must be operated in accordance with operating rules called, in case of nuclear reactors, technical specifications; some of these rules can be deduced from the predictive assessment of operational safety; for example, a nuclear power plant where one of the elements of a given redundant safety system is unavailable: what is the decision to be taken? To interrupt the operation of the installation, with all associated losses, knowing that there are other elements that could fulfill the desired safety function, or to continue the operation, accepting the additional risk of accident, because if an initializing event occurs, there will be a smaller number of means to control its evolution? The predictive assessment of this additional risk will allow such a decision to be made on a rational basis.

8. BIBLIOGRAPHY

[1] Vesely, W.E. et alli, Fault Tree Handbook (NUREG 0492), U.S. Nuclear Regulatory Commission, Washington, USA, 1981.

[2] Comission Electrotechnique Internationale, Liste des Termes de Base, Définitions et Mathématiques Applicables à la Fiabilité, publication 271C, Paris, 1985.

[3] Meyer, P.L., Probabilidades - Aplicações à Estatística, Livros Técnicos e Científicos Editora, Rio de Janeiro, Brazil, 1976.

[4] United States Nuclear Regulatory Commission, PRA Procedures Guide: A Guide to the Performance of Probabilistic Risk Assessment for Nuclear Power Plants, NUREG/CR 2300, US NRC, Washington, USA, 1983.

[5] Llory, M., Villemeur, A. and Portal, R., Les Défaillances de Mode Commun: Identification et Prévention, EDF-DER-HT/13/28/79, Eletricité de France, Clamart, France, 1979.

[6] Villemeur, A., Sûreté de Funcionnement des Systèmes Industriels – Fiabilité, Facteurs Humains ei Informatisation, Éditions Eyrolles, Paris, France, 1988.

[7] Keynes, J.M., Treatise on Probability, Macmillan Books, London, Great-Britain, 1921.

[8] Borel, E., Probabilités et Certitude, Presses Universitaires Françaises (PUF), Paris, 1963.

[9] Raiffa, H., Analyse de la Décision – Introduction aux Choix en Avenir Icertain, Éditions Dunod, Paris, France, 1973.

[10] Mayo D.G. and Hollander, R.D., Acceptable Evidence: Science and Values in Risk Management, Oxford University Press, New York, USA, 1991.

[11] Papoulis, A., Probability, Random Variables and Stochastic   Processes, Mc-Graw-Hill Book Company, New York, USA, 1965.

[12] Kervern, J.Y. and Rubise, P., L'Archipel des Dangers, Economica, Paris, France, 1991.

[13] International Nuclear Safety Advisory Group, Probabilistic Safety Assessment, Safety Series 75-INSAG-6, International Atomic Energy Agency, Vienna, Austria, 1992.

[14] Vigier, M., Méthodes d'Assurance de Qualité-Fiabilité ET d'Expérimentation, Malione Éditeurs, Compiège, France, 1981.

[15]Association Française des Normes Techniques (ANFOR), Gestion de la Qualité (NF50-109), Paris, France, 1986.

[16]Association Française des Normes Techniques (ANFOR), Statistique et Qualité (NF06-501), Paris, France, 1986.

[17] Juram, J.M., Gestion de la Qualité, AFNOR, Collection Normes et techniques, Paris, France 1983.

[18] Wald, A., Statistical Decision Functions, John Wiley and Sons, New York, USA, 1950.

[19] De Finnetti, B., Probability, Induction and Statistics, John Wiley and Sons, New York, USA, 1972.

[20] Batteau, P. and Marciano, J-P., Probabilité et Décision dans l'Incertain, Presses Universitaries Françaises, Paris, France, 1976.

[21] Volkovsyski, L. et ali, Problemas sobre a teoria de Funciones de Variable Compleja, Editorial Mir, Moscow, Russia, 1972.

[22] Lochard, J. et ali, ALARA: from Theory towards Practics, Comission od European Communities, EUR-13795-EN, Luxemburg, 1991.

[23] International Commision of Radiaton Protection, Cost-Benefit Analysis in the Optimization of Radiation Protection, ICRP Publication nº 10, Pergamon Press, Oxford, GB, 1983.

[24] Shrader-Frechette, K.S., Science Policy, Ethics and Economic Methodologt, Reidel Publishing Co., Dordrecht, Germany, 1985.

[25] Munier, B.R., Risk, Decision and Rationality, Reidel Publishing Co., Dordrecht, Germany, 1988.

[26] Farmer, F.R, Reactor Safety and Siting: a Proposed Risk Criterion, in Nuclear Safety nº 8(6), London, GB, 1967.

[27] Layfield, F., The Sizewell Public Inquiry, HMSO, London, 1988.

[28] Health and Safety Executive, The Tolerability of Risks from Nuclear Power Plants, HSMO, London, 1988, reviewed 1992.

[29] Jones Lee, M.W., The Economics of Safety and Physical Risk, Beckwell, Oxford, GB, 1989.

[30] Dreicer, M., External Costs in Nuclear Power Production, Report 132/95, Centre d'Éstudes sur la Protection dans le Domaine Nucleáire, Fontenay-aux-Roses, France, 1995.

[31] Neveu, J., Bases Mathématiques du Calcul des Probabilités, Masson Édituers, Paris, France, 1970.

[32] Feller, W., Na Introduction to Probability Theory and its Application, Wiley and Sons, New York, 1988.

[33] Kaufmann, W., Introduction à la Théorie de Sous-ensembles Flous à l'Usage des Ingénieurs, Masson Éditeurs, Paris, France, 1977.

[34] Ventsel, H., Théorie des Probabilitiés, MIR, Moscow, Russia, 1973.

[35] Tribus, M. Décisions Rationnelles dans l'Incerain, Massan Éditeurs, Paris, France, 1972.

[36] Procaccia, H. et ali, Fiabilité des Équipements et Théorie de la Décision Fréquentielle et Bayesienne, Collection Edf-DER nº 81, Eyrolles, Paris, 1992

[37] Morlat, G., Sur la Théorie de la Décision Appliquée aux Évenements Rares, Nuclear Energy Agency - OECD, Paris, France, 1960.

[38] Linstone, H., and Turoff, H., The Delphi Method, Addison-Wesley Publishers, Chicago, USA, 1975,

[39] Lievens, C., Sécurité des Systèmes, CEPADUES, Paris, France, 1976.

[40] Bureau de Normalization de l'Aéronautique et de l'Espace, Guide d'Élaboration des Objectifs de Securité d'um Système Missile ou Spatial, RG-AERO 701-12, Paris, France, 1987.

[41] Rassmussen, C. et ali, Reactor Safety Study: an Assessment of Accident Risks in US Commercial Nuclear Power Plants, WASH-1400, U.S. Nuclear Regulatory Commission, Washington, USA, 1975.

[42] Rowe, N.D., An Anatomy of Risk, John Wiley and Sons, New York, USA, 1973.

[43] Starr, C., Social Benefit versus Technological Risk: What is our Society Willing to Pay for Safety?, in Science 165-1232, Boston, USA, 1973.

[44] Lowrance, W.W., On Acceptable Risk, Kaufmann, Los Altos, USA, 1076.

[45] Starr, C. et ali, Philosofical Basis for Risk Analysis, inn Ann.Rev.Energy 1-629, Washington, USA, 1976.

[46] Otway, H.J., and Cohen, J.J, Reactor Safety and Design from a Risk View Point, in Nuclear Engineering and Design 13-365, Washington, USA, 1970.

[47]Slovic, P. et ali, The Assessment and Perception of Risk, The Royal Society of London, 34-17, London GB, 1980.

[48] Institut de Protection et Sureté Nucléaire, Rappor du Groupe de l'Observatoire sur les Risques et la Securité, Fontenay aux Roses, France, 1992.

[49] Commission of the European Communities, Summary report on Safety Objectives in Nuclear Power Plants, Report EUR-12273-EN, Luxemburg, 1989.

[50] Boll, M., Les Certitudes du Hasard, Presses Universitaires Françaises, Paris, France, 1958.

[51] Henley, E.J. and Kumamoto, H., Reliability Engineering and Risk Assessment, Prentice-Hall, Englewood Cliffs, USA, 1981.

[52] Le Coq, Les Allocations de Sûreté, course workbook of Mastère de Sûreté da École Centrale, Paris, France, 1989.

[53] Desroches, A., Proposition d'une Méthodologie Préliminarie d'Allocation de Securité Relative aux Différents Configurations d'un Système, in 4ème Séminaire Europeén sur la Securité des Systèmes, Deauville, France, 1986.

[54] McCormick, N.J., Reliability and Risk Analysis, Academic Press, New York, USA, 1981.

[55] Costa Neto, P.L., Estatística, Editora Edgard Bücher, São Paulo, Brazil, 1977.

[56] Brunet-Moret, Y., Statistiques des Rangs, Cahier de l'ORSTOM, Paris, France, 1973.

[57] Dumas de Rauly, D., L'Estimation Statistique, Éditions Gauthier-Villars, Paris, France, 1968.

[58] Gumbel, E.J, Statistics of Extremes, Columbia University Press, Washington, USA, 1958.

[59] Gumbel, E.J., Méthodes Graphiques pour l'Analyse des Débits des Crues, Société Hydrotechnique Française, Paris, 1956.

[60] Hasofer, A.M. and Wang, Z., System Reliability Calculations Using Extreme Value Theory, University of South Wales, Australia, 1985.

[61] Gnedenko, B.V. Sur la Distribution Limite du Terme Maximum d'une Série Aléatoire, Analles Mathematiques, Paris, France, 1943.

[62] Engelund, S. E Rackwitz, R., On Predictive Distribution Functions for the Three Assymtotic Extreme Value Distributions, Elsevier Sciences Publishers, New York, USA, 1992.

[63] Bernier, J.E Veron, R., Sur Quelques Dificultés Rencontrés dans l'Estimation d'un Débit de Crue de Probabilité Donné, Eletricité de France EdF-DER, 1963.

[64] Engelhandt, B., Simple Approximate Distributional Results for Confidence and Tolerance Limits for the Weibull Distribution Based o Maximum Likeboard Estimation, Technometrics vol.23, New York, USA, 1981.

[65] Hammer, W., Handbook of System and Product Safety, Prentice Hall Inc., Englewood Cliffs, USA, 1972.

[66] Power, G.J. et ali, Fault Tree Synthesis for Chemical Processes, AICHE Journal 20 (2), USA, 1974.

[67] Departament of Defence, Military Standard, System Safety Program for Systems and Associated Subsystems and Equipments: Requirements For, MIL-STD-882, Washington, USA, 1975.

[68] Department of Navy, Procedures for Performing a Failure Modes and Effects Analysis, MIL-STD-1629A, Washington, 1975.

[69] Institute of Electrical and Electronic Engineers, Guide for General Principles of reliability Analysis for Nuclear Power Generating Stations Protection Systems, IEE Std 352-1975 (ANSI N41-4-1-1976), USA, 1975.

[70] Commission Eletrotechnique International, Techniques d'Analyse de la Fiabilité des Systemès: Procédures d'Analyse des Modes de Défaillance et de leurs Effects (AMDE), Publication nº 812, Paris, France, 1985.

[71] Society of Aerospace Engineers, Design Analysis Procedure for failure Mode, Effects and Criticality Analysis FMECA, Recommanded Practice ARP 926, USA, 1967.

[72] Pagès, A.E Gondran, M., Fiabilité des Systèmes, Eyrolles, Paris, France, 1981.

[73] Dhillon. B.S. and Singh, C., Engineering Reliability, John Wiley and Sons, New York, USA, 1981.

[74] Anderson, R.T., Reliability Design Handbook, Reliability Design Center NO RDH 376, London GB, 1976.

[75] Pollack, S.L. Decision Tables: Theory and Practice, Wiley Interscience Publishers, New York, USA, 1971.

[76] Relcon Teknik AB, Risk Spectrum, Solnas, Sweden, 1994.

[77] Scientech Inc., REVEAL_W, Rockville, USA, 1994.

[78] Nimal, J-C., Les Simulations de Monte-Carlo, apostila do Cours de Mastère de Sûreté, École Centrale, Paris, France, 1991.

[79] Leory, A., Les Réseaux de Petri, apostila do Cours de Mastère de Sûreté, École Centrale, Paris, France, 1992.

[80] Cullmann, G., Initiation aux Chaînes de Makov: Méthodes et Applications, Masson, Paris, France, 1975.

[81] Alla, H, and Davis, P., Du Grafcet aux Réseaux de Petri, Hermès, Paris, 1989.

[82] Ancellin, C. et ali, Les Méthodes d'Analyse Prévisionnelle de la Fiabilité des Systèmes de Sûreté de Centrales Nucleáires, in 4ème Colloque International de Fiabilité et Maintenabilité, Perros-Guirec, France, may 21-25 of 1994.

[83] Fleming, K.N. et ali, On the Analysis of Dependent Failures in Risk Assessment and Reliability Evaluation, in Nuclear Safety vol 24 nº 5, September-October, 1983.

[84] Fleming, K.N. et ali, A Systematic Procedure for the Incorporation of Common Cause Events into Risk and reliability Models, in Nuclear Engineering and Design, International Post-Conference Seminar SMIRT 8, Bruxels, Belgium, 1985.

[85] Atwood, C.L., User's Guide to Binominal Failure Rate, NUREG-CR-2729, U.S. Nuclear Regulatory Commission, Washington, USA, 1984.

[86] Meslin, T.E Bourgade, E., Common Cause Failure Analysis and Quantification on the Basis o Operating Experience, International Topical Conference on Probabilistic Safety Assessment and Risk Management, Zurich, Switzerland, 1987.

[87] Swain, A.D. and Guttmann, H.E. Handbook of Human Reliability Analysis with Emphasis on Nuclear Power Plant Application, NUREG, /CR-1278, U.S. Nuclear Regulatory Commission, Washington, USA, 1983.

[88] Hannaman, S., Systematic Human Action Reliability Procedure (SHARP), Electric Power Research Institute, EPRI NP-3583, Boston, USA, 1984.

[89] Haugen, E.B., Probabilistic Approaches to Design, Wilwy and Sons, New Yorkk, USA, 1968

[90] Ligeron, J.C., La Fiabilité en Mécanique, Desforges, Paris, France, 1979.

[91] Desrosches, A., Introduction aux Méthodes Resistance-Contrainte, workbook of Cours de Mastère de Sûreté, École Centrale, Paris, France, 1992.

[92] McLeod, J.E. and Rivera, S.S, Métricas para una Programación Confiable, in Anais do VI Congresso Geral de Energia Nuclear, Rio de Janeiro, Brazil, october of 1996.

[93] Favez, B. Les Études de Fiabilité: Utopie ou Realité Industrielle, International RAM Conference, Toronto, Canada, 1993.

[94] International Atomic Energy Agency, The Role of Probabilistic Safety Assessment and Probabilistic Safety Criteria in Nuclear Power Plant Safety, IAEA Safety Series nº 106, Vienna, Austria, 1992.

[95] International Commission of Radiation Protection, Recommendations, ICRP Publication nº 60, Pergamon Press, Oxford, GB, 1991.

[96] Betin, M., Les Effets Biologiques des Rayonnements Ionizants, Eletricité de France, Paris, France, 1991.

[97] International Atomic Energy Agency, Extension of the Principles of Radiation Protection to Sources of Potential Exposures, Safety Series nº 104, Vienna, Austria, 1990.

About the Author

Leonam dos Santos Guimarães graduated in Naval Sciences from Naval School (1980), graduated in Naval and Ocean Engineering from University of São Paulo – USP (1986), has a master's degree in Naval and Ocean Engineering from USP (1991), has a master's degree in Nuclear Engineering from Institut National des Sciences et Techniques Nucléaires – INSTN of the University of Paris XI (1994), has a master's degree in Naval Sciences from Naval War School (1996) and has a PhD in Naval and Ocean Engineering from USP (1999). Currently, he is Director for Planning, Management and Environment of Eletronuclear S.A, member of Standing Advisory Group (SAGNE) to International Atomic Energy Agency (IAEA) Director-General, member of World Nuclear Association (WNA) Board of Management and President of Latin American Section of the American Nuclear Society (LAS/ANS). He was formerly Technical and Commercial Director of Amazonia Azul Defense Technologies S.A. (AMAZUL) and Nuclear Propulsion Program Coordinator at the Technology Center of Navy in São Paulo (CTMSP). On academic positions, he was Full Professor at the School of Administration at Foundation Armando Alvares Penteado (FAAP), Visitant Professor at the Naval and Ocean Engineering Department at Polytechnic School – University of São Paulo (USP), at the Foundation for the Development of Technology and Engineering (FDTE), at the Antonio Carlos Vanzolini Foundation (FCAV) and Adjunct Professor at University of Great ABC (UNIABC). He was also Chief Engineer Officer of the High Sea Tugboat Triunfo and Ocean Sailboat Cisne Branco.
email:leosg@uol.com.br   -   leonam@apeiron.eng.br   -
leonam@eletronuclear.gov.br

Get Published with Frontier India

Do you want to get your book or thesis published? You might even want to republish your book which is currently out of print.. Frontier India Technology as a publisher, distributor and retailer of books, offers a complete range of publishing, editorial, and marketing services that helps you as an author to take his or her book to the reader.

Getting your work published is a wish for many for reasons including profit earning, self-satisfaction, popularity and other good reasons. We will offer you choices based on your needs. Get in touch with us at **frontierindia.org**.

Notes

Notes